CREATION AND/OR EVOLUTION

An Islamic Perspective

T.O. Shanavas

XLIBRIS
Philadephia, PA

To order additional copies of this book, contact:
Xlibris Corporation
1-888-795-4274
www.Xlibris.com
Orders@Xlibris.com
23295

CONTENTS

The Qur'an 29:20

Say: 'Travel in the earth, then behold how Allah originated creation, then He produces a later creation: for Allah has power over all things.'

What you read in these pages is what I heard from those who traveled on the earth to see how Allah originated creation and produce later creation.

ACKNOWLEDGMENTS

I am grateful to my wife, Faizi, and my children, Zaifi and Reema, for their tolerance, understanding, and encouragement despite the fact that work on this book took considerable family time away from them. I have borrowed ideas, words, and phrases from many sources and if I have failed to acknowledge any source, it is not intentional. I am thankful to Professor Samuel Abraham of Siena Heights University in Adrian, Michigan, who provided me with some of the bibliographic materials for the book. I am also obliged to Dr. Maher Hathout of the Islamic Center of Southern California, the late Imam A. M. Khattab of the Islamic Center of Greater Toledo, Professor Syed Hasan of the University of Kansas, and Dr. Nadeem Kutaish, M.D., for their review and critique of the manuscript. Sincere appreciation goes to the Islamic Research Foundation and its president, Professor Ibrahim Syed of the University of Louisville, who played a major part in encouraging me to pursue this work. Without them, this book would never have been completed. I am obliged to Maxxion Technologies, Inc for the illustrations and the cover design. Finally, I am grateful to Laurie Rosin and Nada Najjar for their superior editorial services.

INTRODUCTION

When my son, Zaifi, was in high school, he said, "Dad, you send me to the best school around our home to study science. You send me to the Islamic Center to study the Qur'an. Science says that human beings evolved from the world of apes, but the Islamic Center teaches us that humans were initially created in heaven and came to earth fully formed. What is the truth?"

I could not answer him immediately. His curiosity motivated me to study the concept of Creation in Islam and led to the writing of this book. Through my studies, my reply to my son—and to all who wish to teach or study Creation, of any faith, scientist or theologian—became clear: Science and religion need not be competing ideologies.

This book has several objectives. First it describes aspects of the Islamic view of Creation and seeks to arouse interest in the integration of modern knowledge with religion.

Second, it brings to light important historical facts about the theories of evolution and uniformitarianism (a subject treated in detail in chapter 4) that have been dismissed or forgotten. Although Western scientists claim, "The morality of science is reporting the truth,"[1] they recount the history of evolution starting only from Lamarck and Darwin. They ignore the scientific contributions of Muslim scholars or present them as gifts from the Greeks to humanity. I seek, therefore, to set the historical record straight, for surely an ethical interpretation of scientific evidence requires accurate reporting of the truth as a whole.

Third, the book introduces an additional dimension into the evolution-creation debate over the policy of teaching creationism in the schools. In the United States, Christian fundamentalists are

11

pressuring every level of administration of our educational systems, Congress, and the courts to include their version of creationism in the school curriculum. This book serves to inform fundamentalists and lawmakers of other creationist beliefs that exist among the citizens of the United States of America.

Many Muslims, particularly Adnan Oktar, known as Harun Yahya, lavishly promote fundamentalist Christian creationism among Muslims. Harun Yahya, in cooperation with fundamentalist Christians, is spearheading a massive campaign against the theory of evolution. Many Muslim magazines and newspapers in the United States and other parts of the world are helping Yahya poison Muslim minds against science. Ironically, this practice often provokes an unfortunate backlash, generating skepticism about Islam.

The process by which this occurs is insidious. People like Harun Yahya, using quotes from scientific journals but applying them out of context, mislead those Muslims who are scientifically unschooled into believing the fundamentalist Christian doctrine of the so-called scientific creationism.[2] Following the modus operandi of the fundamentalist Christian organization Institute for Creation Research (ICR), Yahya uses pseudoscience to promote his interpretation of the Qur'an. The references he cites in his book, if read in their entirety, usually accept and defend evolution. But he routinely selects just one sentence from an article, a line that can be construed to support his arguments, and uses it as a scientific reference. Like the ICR, he distorts single news items from popular journals to "prove" his conclusion. He conveniently ignores the rest of the article or other articles in the same issue of the journal that support evolution.

Such pseudoscience promulgated by Muslim cohorts of fundamentalist Christians lead young Muslims studying science at American universities into distrust or rejection of Islam as a whole because they soon discover the flaws and half-truths in this imported "scientific" creationism.

Sadly, the young Muslim men and women studying in Western universities get little guidance from the imams, the religious guides of each Muslim community, at their mosques. Whether in the

West or the East, most imams have little knowledge of science. When asked about human evolution, they say it is a theory, but they cannot render a scientific definition of the word. Imams in general are incapable of giving informed, credible answers to our budding anatomists, embryologists, paleontologists, molecular biologists, physicists, chemists, and physicians.

Many Muslim parents living in the West have scientific training, which they use in their daily life. But they, too, can be poor guides for inquisitive young Muslim minds. They tell their children that scientific theories may be proved wrong at a later time. They ask these intelligent young men and women to listen to their imams and believe what they say. Simultaneously they not only encourage but also often pressure their children to major in scientific fields. What an amazing contradiction—lauding the study of science but rejecting its discoveries!

I hope that by means of this book Muslims will remind themselves of their contributions to the theory of evolution predating Charles Darwin and stop propagating the Christian fundamentalist doctrine of creationism in their community. I hope also that Muslims will join hands with mainstream scientists to prevent the corruption of public-school science curricula with pseudoscience. And finally, I hope it may in some measure rescue young Muslim science students from the tragic quandary in which they now find themselves.

Imams of *madrasas* (Arabic for schools)—who turn innocent Muslims into monstrous killers, as we have seen in the September 11 incident—preach that Adam was created ex nihilo in heaven, the eternal Garden of Final Return, without any connection to other creatures of the animal kingdom. This book will expose the dichotomy between the imams' *kutuba* (sermons in mosques) and simple, literal readings of the Qur'an regarding Creation. I hope thereby to unveil the ignorance of these imams, so all Muslims might see and understand the true teachings of their faith and so young Muslims will not trust the imams who distort Islam.

I have pursued the study of Creation for both religious and personal purposes. The Qur'an teaches Muslims that those who

are pious shall worship God through the "signs in the Heavens
and the earth and in [themselves]." The "signs of God" to which
the Qur'an constantly refers are primarily the phenomena of the
natural world:

> Behold! In the creation of heavens and the earth, and
> alternations of night and day,—There are indeed Signs
> for men of understanding,—Men who celebrate the
> praise of God, standing, sitting, and lying down on their
> sides, and contemplating the (wonders of) creation in
> the heavens and the earth: 'Our Lord! Not for naught
> hast Thou created (all) this! Glory to Thee! Give us
> salvation from the penalty of fire.' (Qur'an 3:190-191)[3]

Finally, I consider my effort to illuminate the Muslim portrait
of creation as an act of worship (*ibadath*). The prophet's tradition
states: "Allah said: 'I was a hidden treasure; I wanted to be known;
therefore I created the world so that I would be known.'"[4] As the
tradition suggests, Islam does not clearly delineate between the
natural and supernatural. The physical universe is an integral part
of the Muslim's religious universe. Humanity has to explore the
universe and its laws to know God. Hence, understanding the
spiritual significance of an active nature (*kiyan*) is an important
stepping stone on the road to Islamic spirituality. The sun, moon,
and stars are not mere heavenly bodies but religious realities that
participate in the Islamic vision of the universe.[5] In the Qur'an,
God even takes them as witness on the Last Day: "So, I call the
receding stars to witness." (Qur'an 81:15)[6] Both the Qur'an and nature
urge human beings to worship God: "We shall show them our portents
on the horizon and within themselves until it will be manifested
unto them that it is the truth . . . ." (Qur'an 41:53)[7] Therefore, a
contemplative attitude toward nature is a form of worship.

The main theme of this work is an inquiry into the creation of
humanity and the physical universe. According to scripture, God
created other, nonphysical beings as well—angels and jinn. The
Qur'an states that God created angels from light; traditionally they

are described as almost robotic, though luminous, servants of Allah with no free will. Jinn, by contrast, are created from fire-smoke, and unlike angels they are endowed with free will. This book does not discuss the creation of supernatural beings, because they are imperceptible and therefore cannot be studied through physical observation.

Even though I have disagreed with many Muslims, Christians, and Jews on some of their interpretations of the story of Creation, I do not claim to be more virtuous than them or other religious people. I share their belief that existence emanated from God, but I do not believe in a sectarian God. If God is as just as all religions claim, God cannot be sectarian. On the Day of Judgment, a just God's reward or penalty to human beings cannot be based on which group they were born into or chose to follow but solely on their righteous deeds and faith. The following verse from the Qur'an confirms this view:

> " . . . For every soul will receive its reward by measure of its endeavor." (Qur'an 20:15)[8]

In this verse, the phrase "every soul" negates the *contemporary* Muslim claim of exclusivity for God's love and reward, unless Muslims make a parallel assertion that Christians, Jews, and others do not have souls and Muslims are the only ones with souls.

The Qur'an elaborates further:

> It is not your [Muslims'] whims, nor by the whims of the People of the Books [Jews and Christians]. Whoever does misdeeds shall be recompensed for it, and shall not find for him, apart from God, neither patron, nor helper. And whoever, whether male or female, does righteous deeds while believing, shall enter the garden and not be wronged a jolt. (Qur'an 4:123-124)[9]

> Those who believe (in the Qur'an), and those who follow the Jewish (scriptures), and the Christians and the

Sabians—any who believe in God and the Last Day and work righteousness shall have their reward with their Lord: on them shall be no fear, nor shall they grieve. (Qur'an 2:62)[10]

Yet they are not all alike; some of the People of the Book (Christians and Jews) are a nation upstanding, that recite God's signs in the watches of the night, bowing themselves, believing in God and in the Last Day, bidding to honour and forbidding dishonour, vying one with the other in good works; those are of the righteous. And whatsoever good you do, you shall not be denied the just reward of it; God knows the god-fearing. (Qur'an 3:113-115)[11]

The Qur'an even goes further to counsel the believers:

And you will find those nearer in affection to those who believe (i.e., to the Muslims) who say, 'We are Christians', that is because some of them are priests and monks, and because they are not haughty. (Qur'an 5:82)[12]

Finally, the Qur'an describes the Torah and the New Testament as books of guidance and light. It even states that Allah allows the people to practice different religions and to compete with each other in doing good deeds. I have added the underscoring for emphasis:

We sent down the *Torah which contains guidance and light* . . . . Later, in the train (of Prophets), We sent Jesus, son of Mary, confirming the Torah which had been (sent down) before him, and gave him the *Gospel containing guidance and light* . . . . And to you [O Muhammad] We have revealed the book [Koran] containing truth, confirming the earlier revelations . . . . To each of you [Muhammed] We have given a law, a way and a pattern of life. *If God had pleased He could surely have made you one people (professing one faith).* But He wished to try you and test you by that which He gave you. So try to excel in good deeds. To Him will you all return in the end, when He will tell you of what you were at variance. (Qur'an: 5:44-48)[13]

How true these verses ring! I have been acquainted with Jews, Christians, Muslims, Hindus, and people from other religious denominations. Many are kind, compassionate, and willing to sacrifice worldly riches for the benefit of all of humanity. Many are less envious and more righteous than I am. Albert Sabin, the Jew who invented the oral polio vaccine; Mahatma Gandhi, the Hindu and man of peace; and Alexander Fleming, the Christian inventor of penicillin, were great men who committed themselves to the welfare of human beings. Father Damien, who served the lepers; former US president Jimmy Carter, a Christian peacemaker who builds homes for the homeless; and Mother Theresa, a Catholic nun who helped and comforted the poor in the slums of Calcutta, are also fine examples of individuals who dedicated their lives to the service of humanity and worshipped God through selfless service. The Bible inspired Mother Theresa, Jimmy Carter, Father Damien, and millions of Christians to dedicate their life to relieve the pain and suffering of all people. They vindicate the Qur'anic dictum that guidance and light can be found in the Holy Bible. I may disagree with some of their religious beliefs, but in light of the Qur'anic verses cited above, it would be absurd for me to imagine that they would not receive more compassionate and merciful treatment from Allah than I would as a Muslim.

Thus, the conclusions I draw in this work about certain *contemporary* Judeo-Christian-Islamic interpretations of Creation are not an expression of dislike toward these traditions. They constitute a rational, mostly literal, interpretation of the Muslim Holy Book.

I hope that this book will be useful to scientists who are ignorant of the extensive history of Muslim thought on evolution and to non-Muslims who have little knowledge about the Muslim interpretation of Creation during the classical period of Islamic history. Further, I hope this book creates a sense of curiosity in the mind of Muslims who may have lacked the will to question the irrational pronouncements made by some of today's religious scholars.

Finally and most importantly, I wrote this book with the intention of fostering understanding among all peoples of the earth; I fervently hope that it will not be used by anyone to advance the cause of religious intolerance. Otherwise, I will have failed in my endeavor. I believe that human beings are created with pluralistic and multicultural qualities. We are not here to alter God's creation by force or hateful distortions. Rather, we ought to appreciate the beauty and versatility of this world and affirm the principles that enhance harmony and tolerance among the human race. The Qur'an supports this harmony in chapter 49, verse 13:

> O mankind! We created you from a single (pair) of female and male, and made you into nations and tribes, that ye may know each other (not that ye may despise each other). Verily the most honoured of you in the sight of God is (He who is) the most righteous of you. And God has full knowledge and is well acquainted (with all things). (Qur'an 49:13) [14]

A note on translations of the Qur'an:

> Please note that throughout the text I use literal translations of the Qur'anic verses, as is common in Muslim scholarship. The translations of the Qur'anic verses quoted herein should be viewed as the opinion of the translator, however.

Muslims believe that it is humanly impossible to translate the Qur'an or to reproduce its majestic beauty, indescribable rhythm, and rhetoric. Muslims believe that there is only one Qur'an, which has been preserved in its original form in its original Arabic language. Thus, any translation or commentary in any other language is considered the opinion of the translator or commentator.

The word *Allah* is an excellent example by which to illustrate the problems entailed in translating the Qur'an into English. Allah is the Arabic term for God, the Supreme Creator of all that exists. There is no plural for the word Allah in Arabic. Allah has no partners, sons, or daughters. Allah is self-sufficient, and no deities precede or exceed "him." Even though the male gender is assigned in English translations for Allah, Allah is neither male nor female.

Because it is nearly impossible for an author to extricate himself from his work, my vision of the world has undoubtedly influenced the thoughts expressed in this book. I have applied every effort to remain close to the literal text of the scriptures in drawing an accurate picture of the concept of Creation.

CHAPTER 1

Inshah Allah (Allah So Willing)

The Islamic Metaphysics of the Future

In the life of devout Muslims, a day never passes without using the Arabic phrase *Inshah Allah* (God So Willing) at the end of any conversation about future events. Without an understanding of the meaning of this phrase, we cannot begin to comprehend God's relationship with His creatures, the concept of Creation, and the role of free will. Nor can we offer a rational, internally consistent argument against the materialist's exclusion of God in the evolution of life and the universe.

Materialist scientists argue that biological evolution is an "inherently mindless purposeless process."[1] They preach that impersonal laws rule the universe, and atoms are at work in the operation of life. Biologist and atheist Richard Dawkins insists that contingency and natural selection, operating over a long period of time, account for evolution. Dawkins assumes that blind forces of physics, chemistry, and natural selection are sufficient to explain the origin and expansion of life.[2 & 3] He asserts that the unfolding of life is the result of selfish desires of genes to increase their opportunities for survival and reproduction.

Similar opinions prevail among other practitioners and admirers of science who argue that there is no reason to include God in the evolution of life. One extremist states, " . . . materialism is absolute [and] we cannot allow a Divine Foot in the door."[4]

Such fervency stems from an unshakable, unwavering faith in the Law of Causality, which most people acknowledge and which states that a given cause always produces the same effect. Gravity always pulls an apple down to the earth; spring season melts snow; drought brings destruction of crops. Chemical reactions in any organism, amoeba or human, are explainable by the same laws of physics and chemistry that govern the universe.

Based on causality, scientists maintain that the future is predetermined and can be predictable through accurate knowledge of past causes. The laws of nature, they argue, are invariant, and scientific observation reveals the past as the product of those laws. Any natural event that departs from the anticipated effect of a uniform cause is classified as an "accident." However, scientists' predictions based on observation of matter and invariant laws of nature are limited by their own earlier conclusions and experiences.

To gather data, scientists peer into nature, from atoms to stars, amoebas to humankind, fungi to maple trees, and any other phenomena of our universe. Science has categorized the collected data, defining disciplines such as paleontology, comparative anatomy, biogeography, embryology, molecular genetics, and so on. The materialists' claim that the unfolding of life is a "purposeless, mindless process" is based upon inferences from separately catalogued extrapolations of past experiences. John F. Haught, professor of Theology at Georgetown University, calls such materialistic metaphysics "metaphysics of the past."[5]

Haught's outstanding treatise *God After Darwin* helped me develop a better understanding of Islam and the concept of Creation described in the Qur'an.[6] I find his mode of thought to be as much Islamic as his belief system, so I shall apply his metaphysics to an Islamic context.

With unflinching and, on the surface, contradictory faith that God created everything, many Muslims also believe that the past determines the future. In the Islamic universe, unlike that catalogued by materialist science, the past and the present are not the creators of the future, nor are humans or any other creatures, because " . . . Allah is the creator of everything . . . ." (Qur'an 13:16)[7] Even creations that we claim as our own emanate from Allah. The Qur'an states, "And God created you and what you make." (Qur'an 37:96)[8] Allah created everything—computers, airplanes, cars, and even the atom.

A few important questions Muslims might ask are, if God is the creator of what humans make, is God not also the creator of our good and bad deeds? Why should there be reward for the pious and retribution for the impious if Allah is the source of our actions? If Allah has sealed the heart of disbelievers from receiving divine guidance (Qur'an 2:7), why does God hold us accountable for our actions on the Day of Judgment? The answer to these questions lies in Islamic metaphysics of the future.

The experienced past is irretrievable; the present is only a fleeting moment that we cannot hold. On the other hand, we experience the continuous coming of the future. Future in the material world does not exist until it is created. Future is not simply the birth of a moment. *Future* means the yet-to-be-born or created moment packed with contrasting or diametrically opposite possibilities. Each moment brings each of us hope or fear, success or failure, pain or pleasure, routine or surprise events. Islamic faith decrees that Allah is the source of all information. Therefore, Muslims pray, "My Lord, augment me in knowledge." (Qur'an 20:114)[9]

Islamic teachings regarding the coming of future events are grounded in the phrase *Inshah Allah* (God so willing) and the verse, "And never say about anything, 'Behold, I shall do this tomorrow,' without [adding] 'if God so wills.'" (Qur'an 18:23).[10] Muslims say Inshah Allah after every statement pertaining to the future, even for simple tasks such as meeting a friend at 4 P.M. tomorrow. Muslims believe that the future is not simply born without cause. It occurs only when and if Allah creates it as information within arriving moments and if His creatures act upon the information to actualize it into visible realities in their material world. Our planning and our desires may or may not be what Allah is going to present to us in our future. Allah states: " . . . and they contrived, and God contrived, but God is the greatest of contrivers." (Qur'an 3: 54)[11]

All creatures participate in actualizing the possibilities contained within future moments into visible monuments of God. The present is the pivotal moment between past and future. Allah tests us by asking us to make moral choices of possibilities—the good, bad, and neutral, the moral and immoral—that are contained in each approaching moment only as information. These choices have no negative or positive charge in the material world until creatures actualize the possibilities into material-world realities. Thus we are necessary participants in the ongoing process of creation. Otherwise the stream of future moments would remain only as idle information and possibilities. Allah calls humans to account for our act of choosing from the alternatives within the arriving future moments and also for rejecting information that appears logically correct to each individual.

Each moment is different for each individual, and the lessons of the past and present are tools that can help us make the right choices. Allah revealed His words to the prophets and provided holy books to both humans and *jinns* (beings who inhabit a different plane of existence). These scriptures explain what is right or wrong, moral or immoral, preferred or not preferred, and rewarding or punishable. Allah narrates stories of the past in the Qur'an in order to help humans and jinn—both of whom He has graced with limited free will—use lessons from history in conjunction with revelations to make the right choices.

Living creatures have the freedom to actualize any possibilities contained as information in each moment of future arriving from God. To label the above belief, we shall borrow John F. Haught's phrase "metaphysics of the future"[12] and modify it to read "Islamic metaphysics of the future." If a future moment arrives lacking in novel possibilities, humans and other creatures cannot change their present condition, which then becomes stagnant and may remain so for an unlimited period of time. Even when Allah sends our way moments with novel possibilities, we will remain unchanged if we do not accept His offerings and revelations.

In Islamic metaphysics of the future, the universe is always within Allah's providence. Therefore, Allah is the creator of all things and the one who brings the possibilities of each moment in the form of information. Nothing comes into existence without the information about it being initially available. Similarly, inventors and technicians provide the information they receive to various sectors of society—such as politicians, heads of corporations, and others—guiding them to the scientific basis for managing their vocations. Those who understand the information can then actualize it into cars, airplanes, nations, and so on.

I like to use the metaphor of the factory worker to illustrate our relationship to God. While the ordinary assembly-line worker can choose the manufacturing plant in which he wishes to be employed, factory workers have no freedom to manufacture any products of their choice; they must assemble a product using components coming through the conveyer belt of the factory.

The physical and spiritual universe is the manufacturing plant of Islam owned by Allah, the supreme Scientist and Technician. Here creatures at large, and humans in particular, are like assembly-line workers. The chain of flowing moments of the coming future is the conveyer belt, which delivers the raw materials (possibilities as information) necessary for the making of many products. In this factory the worker is free to select any of the components (possibilities) from the conveyer belt (arriving future) and actualize those possibilities into visible monuments of Allah's creation. If there is no flow of information from scientists and technicians, the assembly-line worker is unable to produce anything. Even the factory would not exist. Likewise, humans or any other creatures cannot produce or act upon the world until the future moment arrives with possibilities as information from Allah.

Belief and disbelief in God also come as possibilities in the flow of moments from the future. If human beings choose and accept disbelief in God, their minds become unreceptive to divine revelations until they are willing to give up their disbelief. Consequently individuals opt to receive reward or retribution in the hereafter universe (al-Akhirah) based upon their earthy choices.

The universe from Big Bang to Big Crunch is a maze. On both ends there is singularity where all matter is condensed into a mathematical point. The maze is made out of alleys, roads, highways and byways that lead to different futures for the universe and its components before it faces the Big Crunch (the Last Day). (Figure 1-1). Allah has already mapped out (created) all the possible and available futures that we can choose, but it is still up to humans and other components of the universe, day by day or moment by moment, to decide for themselves which alleys or roads or highways to step into. Allah, the Merciful and Benevolent, does not interfere or force us into making choices and voluntarily limits His Absolute Power as stated in the Qur'anic verse: "And had your Lord willed, whoever in the earth would have believed altogether. Will you then coerce the people to become believers?" (Qur'an 10:99)[13] God knows that free will would be nonexistent for His creatures without limitation of His omnipotence and omniscience. Creatures would not be able to choose when the future arrives with possibilities from God without voluntary self-control of His power and absolute knowledge. Therefore, Al-Rahman (The Beneficient) and Al-Rahim (The Merciful) God set a voluntarily self-imposed limitation on His omniscience and omnipotence to create free will for His creatures. Because the self-imposed limitation is voluntary it does not imply any inherent limitation in Allah's ultimate power and omniscience. At the same time we are free to choose and actualize any of the worldly possibilities available to us—atomic power, computer technology, biological engineering—but our future is limited by possibilities that God has in store for us. In another words, Allah knows all available futures for creatures, but in order to create free will for His creatures Allah, being the Most Merciful and Most Benevolent, voluntarily opted not to know which future path that His creatures would choose to step into until it is done.

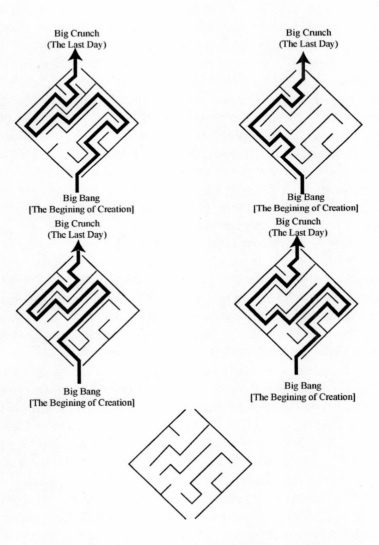

Figure 1-1. Evolution of the Universe. An evolutionary maze of the universe with four different travel routes from Big Bang (The Beginning of the Creation) to the Big Crunch (The Last Day).

In mandatory daily prayers, Muslims recite the opening chapter of the Qur'an, *al-Fatihah*. "Guide us (O Lord) to the path that is straight, the path of those You have blessed, not of those who have earned Your anger, nor of those who have gone astray."[14] These verses imply to many Muslim and non-Muslim minds that Allah does lead some of us astray. Based upon the Islamic metaphysics of the future, Allah presents contrasting and diametrically opposite possibilities within arriving moments of future, and He also states in his revealed books which possibilities please Him or draw His anger. We are thus graced with guidance as to how to travel in the righteous path, by choosing righteous possibilities from the arriving moments of future. Similarly we are offered guidance about which creaturely choices from the possibilities in the arriving moments of future lead us astray.

Even though Allah is the source for the good and sinful possibilities as well as the guidance for selection or rejection of possibilities that are righteous or evil respectively, the possibilities remain as idle information until we actualize any of the information from the arriving future into the reality of the material world. Allah does not interfere in the free choice of His creatures from among the available possibilities within arriving future moments.

The sinners' own confession, as repeatedly mentioned in the Qur'an, is that their great leaders or the Devil misled them. Not once do they put forward the excuse that God Himself misled them: "And they (shall) say, 'Our Lord, we obeyed our masters and our notables, but they led us astray from the path.'"[15] (Qur'an 33:67) "And those who disbelieved shall say, 'Our Lord, show us those who led us perverse of men and jinn, and we shall put them under our feet that they may be at the bottom.'" (Qur'an 41:29)[16]

If God had really led them astray, the sinners' best excuse on the Day of Judgment would have been that they did not deserve to be punished because God Himself led them astray. Therefore, based upon the Islamic metaphysics of the future, the phrases, "God misled" and "God guided," do not mean that God preordained a particular human conduct. But it means that God is the source of all good and bad information. All bits of information from God remain as idle possibilities until creatures translate these into visible realities of the material world.

Based upon the far-reaching meaning of the phrase *Inshah Allah* and the concept of "metaphysics of the future," accidents or contingencies are novelties coming from Allah, even though these novelties appear random to a human mind fixed in the materialist "metaphysics of the past." For example, God created for Hind d. Utba, one of the foremost enemies of the Prophet Mohammed (peace be upon him), the possibility of joining the distinguished company of the Prophet (pbuh), as Hamzah did, or the decision to kill and mutilate Hamzah. Allah did not compel Hind to choose any one of the paths. Allah created good and bad choices in Hind's arriving future, and she decided to actualize the bad choice to kill and cannibalize.[17]

The Qur'an supports the above when it states: "Surely God does not wrong anyone, they wrong themselves." (Qur'an 10:44)[18] No conflict arises, therefore, between the belief that Allah is the creator of the world and the belief that human beings and other creatures are endowed with limited free will within the bounds of possibilities that come through the flow of moments from the future.

Such a construction of the universe and limited free will for His creatures along with merciful voluntary limitation of His power and omniscience distances God as a tyrant. If we take a global picture of all earthly creatures, we witness a universe filled with a mixture of pain and suffering as well as happiness and peace. Christian theologian and biochemist Arthur Peacocke describes the universe as a musical play wherein actors (God's creatures), at their individual hierarchical ranks, freely choose their roles from possible scenarios presented to them by the choreographer (God).[19] Chance and unpredictability are inevitable and are, in fact, built into such an atmosphere. In real life, however, the pain and suffering we experience as a result of the free choices of all creatures for their selfish benefits are God's way of testing, perfecting, and preparing us for the eternal world of absolute happiness and peace.

Sufi poet Jalaluddin Rumi describes this universe as a

battlefield where atom struggles with atom like faith against infidelity. In this struggle some benefit and others suffer. The Islamic metaphysics of the future blends chance, unpredictability, and creatures with limited free will to form our universe within the context of the existence of a Most Compassionate, Most Just, Omnipotent, Islamic God known as Allah. In this universe Allah proposes and humankind and other creatures together dispose. Allah's rewards or retribution in the Hereafter world is based upon our intention and selection of the righteous or the evil information within the messenger-moments called *future* and its actualization into visible monuments God's creation in the material world. This conclusion is supported by the following verse: "Whoso interveneth in a good cause will have the reward thereof; and whoso interveneth in an evil cause will bear the consequence thereof. Allah overseeth all things." (4:85)[20]

CHAPTER 2

The Concept of Time:

New Physics and the Qur'an

"What is time?" I asked my seventh-grade teacher.

Unfortunately, my question enraged him because he thought I had made a serious grammatical error. He advised me that the correct way to inquire was to ask, "What is the time?" In those days corporal punishment was applied in schools, so I was afraid to explain to him what I meant.

Time passed. Days grew into months, months into years, and years into decades, but my curiosity did not change. I knew that time is what a clock measures, but I continued to ponder the question "What is it that the clock measures?"

This question and many more became the focus of my attention when I started to study the Qur'an, the Holy Scripture of the Muslims. A Muslim cannot ignore nature and its laws. Why not? Sayyed Hossein Nasr, an eminent Islamic scholar, answers that question very eloquently:

> The Qur'an depicts nature as being ultimately a
> theophany which both veils and reveals God. The forms
> of nature are so many "mosques" which hide various
> Divine Qualities while also revealing these same Qualities
> for those whose inner eye has not become blinded by
> the concupiscent ego and centripetal tendencies of the
> passionate soul.[1]

The Qur'an and the physical universe are "twin manifestations of the divine act of Self-revelation."[2] Viewed as a text, the universe is a "written scroll" (Qur'an 21:104)[3] with information that must be read according to its meaning. The Qur'an is its counterpart, a text in human vernacular that bids us to explore and coexist with the universe without damaging it.

The Qur'anic verses are called *ayath* (verses), as are the phenomena of nature. The earth, sky, mountains, and stars, the oceans and the ships that float upon it, and all the living creatures in this universe are ayaths (verses). Both the Qur'an and the phenomena of nature are direct communications from God to mankind.

According to the Qur'an, the duty of every believer is to use his or her God-given intelligence to understand the universe and its elements, with the aim of reinforcing his or her faith in God and in His power. The Qur'an draws our attention to the fact that the universe, as a whole, is here for human beings to explore. Through such exploration, humanity can understand the nature of matter, the physical laws that govern it, and ultimately God Himself. The following verse supports this belief:

> And in the earth are signs [ayath] for those who have
> firm faith, and in your own selves. Do you not discern?
> (Qur'an 51:20-21)[4]

Moreover, in the ayaths quoted in the introduction (Qur'an 3:190-193)[5], the Qur'an commands humanity to ponder the wonders of creation in the heavens and the earth. The verses warn Muslims about severe retribution for those who shy away from studying the ayaths contained in the universe. Those who do not seek rational knowledge by observation, listening, and hearing are depicted as the lowest of beasts in the sight of God.

> The worst animals before God are the deaf, the dumb,
> and those who do not use their reason. (Qur'an 8:22)[6]

According to the Prophet (peace be upon him), the Qur'an does not benefit anyone who does not have the basic knowledge to understand it. The Hadith (sayings attributed to the Prophet Mohammed) assert:

> To rise up at dawn and learn a section of knowledge is better than to pray one hundred *rakah* [repetitions of prayer with movement]; it is better than the world and its contents; knowledge is a treasure-house and its key is inquiry. So inquire, there are rewards for four persons: the inquirer, the learned man, the audience and their lovers; to be present in an assembly with a learned man is better than praying one thousand rakahs. The messenger was asked, "O Messenger of God! Is it better than the reading of the Qur'an?" He replied, "What benefit can the Qur'an give except through the knowledge?"[7]

Based on the above verses from the Qur'an and the Hadith, the religious duty of every Muslim is to understand all creations of God so that he or she can comprehend God. Time, being one of the created phenomena of the universe, is among the subjects of contemplation and analysis by Muslims.

Here is another reason to study the nature of time. Muslims, without ever asking the nature of time, believe in the suddenness by which the universe and mankind were created within our earthly time frame. This belief has led them to the rejection of paleontological, biochemical, and other scientific data that biologists use to infer the evolutionary birth of life, plants, animals, and human beings.

When Muslim students of science point out the dichotomy between science and the contemporary Muslim teaching in their mosques, many imams and Muslim parents tell students that reason cannot be trusted and that the human intellect is unreliable—but they forget they are using their intellect to say so! Why should anybody trust their opinion on any subject when they distrust the capabilities of their mind?

Human beings must exercise their intellect to choose imams' religion over other religions or vice versa. Otherwise, how can one determine who has chosen the true path of God? Many would argue that the choice of one's spiritual path is best made in the heart, not in the mind, and that our limited mind is no more capable of determining the "true path of God" than of grasping God's eternity or omniscience. Many Muslims argue that our heart can experience this, but not in the form of thought. Here Muslims are forgetting the Qur'anic verse equating those who do not use their intellect to instinctive animals:

> They have hearts [hearts are often used in the Qur'an as synonymous with minds], but they do not comprehend with them; and they have eyes, but they do not perceive with them; they have ears, but they do not hear with them. They are like cattle; nay, they are even more perverse. Those are they, the neglectful. (Qur'an 7:179)[8]

Hence, use of intellect and human understanding of the world is very important, as is further illustrated by these sayings of the Prophet (peace be upon him):

> God hath not created anything better than reason, or anything more perfect, or beautiful than reason; the benefits which God giveth are on its account; and understanding is by it, and God's wrath is caused by disregard of it.[9]

> It is not a sixth or a tenth of a man's devotion which is acceptable to God, but only such portions thereof as he offereth with understanding and true devotional spirit; Verily, a man hath performed prayers, fasts, charity, pilgrimage and all other good works; but he will not be rewarded except by the proportion of his understanding.[10]

At the same time Allah is merciful to his limited creatures and their limited intellect. God rewards those who genuinely search for the truth even though they may have drawn wrong conclusions from their search. This is reflected in the following saying of the Prophet (peace be upon him):

> When a judge gives a decision, having tried his best to decide correctly and is right, there are two rewards for him; and if he gave a judgment after having tried his best to arrive at the correct decision but erred, there is one reward for him.[11]

The genuine search for truth requires a will to doubt. Such skepticism is not anti-Islamic but a fundamental teaching of the Qur'an. When Prophet Abraham (peace be upon him) said to Allah, "My Lord, show me how you raise the dead," God asked, "Have you not believed?" He [Abraham] said, "Yea, but to make my heart well assured." (Qur'an 2: 260)[12] The Prophet's (pbuh) commentary to this verse is, "We have more rights to be in doubt than Abraham (pbuh) when he said, 'My Lord! Show me how thou wilt raise the dead.'"[13] This saying of the Prophet (pbuh) and the Qur'anic verse make the point that even though human intellect is limited, the use of our intellect is as important to the search for the truth as revelation.

Human beings since the beginning of civilization have pondered the concept of time. The unyielding irreversibility of the passage of time is borne in human beings by the certainty of death. Unlike other life forms, we know that our life could end at any moment, and even if we attain all our earthly expectations, our success is inevitably followed by eventual decay and, in due time, death.

The notion of time in ancient civilizations was much different from that of today. People did not perceive time as a linear continuum that stretches into a continuous future. They pictured it as cyclical in nature and therefore believed that historical events also followed a cyclical pattern.[14]

The Greek philosophers, including the Orphics, Pythagoreans, and Plato, held the view that people are reborn in the flow of time. They taught that our perception of one lifetime per human being is an illusion due to the loss of memory about past lives upon rebirth. Some Greek philosophers, such as Pythagoras and Empedocles, were said to be able to recollect their previous lives. Buddha also recollected all his previous lives. Hindus believe that people repeat cycles of birth and death until they break the succession through their vigorous ascetic performance. On observing repetitive earthly phenomena such as rotation of day and night or the four seasons, the ancients inferred that time and everything else, including human birth and death, is also cyclic.[15]

Pre-Islamic Arab pagans considered time a deity (*ad-dahr*) that exists from eternity to eternity and dispenses good and ill fortune to mankind.[16]

The monotheistic religions—Judaism, Christianity, and Islam—led the way to the concept of linear time. The chronology in the Jewish and Christian scriptures implies that the universe was created in 4,004 BC.[17] The Muslim scripture does not reveal a specific time for its creation, even though some Westerners hold that the Muslims also believe in a young universe. For example, Professor Stephen W. Hawking, one of the greatest minds of the twentieth century—hailed as an "equal of Einstein" by *Time* magazine—states: "According to a number of early cosmologies and the Jewish/Christian/Muslim tradition, the universe started at a finite and not very distant time in the past."[18] Hawking accurately reads the three monotheistic religions when he declares that the universe was created at a finite time, but he errs in assuming that the Muslim estimation of the age of the universe is the same as that held by the Jewish and Christian traditions.

Muslims believe that the universe was created in a finite distant past, but they also believe that it will be destroyed sometime in the future. (Chapter 3 will discuss this in greater detail.). The agreement among the three monotheistic religions on the concept of resurrection, whereby humans live from eternity to eternity, is based on a linear rather than a cyclical chronology.

The invention of the clock in the thirteenth century reinforced the notion of linear time. Since then time has been broken down into units of hours, minutes, and seconds and conceived of as something that progresses from the past to the future.

Measuring time affected the lives of people of all faiths. For example, during the fourteenth century, workers in the cities set up specific times for the beginning and the close of a workday.[19]

Isaac Newton believed that he knew what time was. At the beginning of his *Principia Mathematica,* he wrote: "Absolute, true and mathematical time, of itself, and from its nature, flows equably without relation to anything external."[20] This was the dominant belief among scientists until the twentieth century.

Today we know that Newton was mistaken in several respects. Time is not absolute or universal but relative. As physicist Paul Davies observes:

> Einstein demonstrated that time is in fact elastic and can be stretched and strung by motion. Each observer carries around his own personal scale of time, which does not generally agree with anybody else's. Our individual perception of time does not appear distorted to us, but for observers who move in different time frames than ours, we seem to be out of step with their time.[21]

The time between two events at two different locations is greater for the earthbound observer than for a space-traveling observer. This is called time dilation. The Theory of Relativity states that one day for a space traveler, depending on his velocity, can equal a few years or more for an earthbound person. This startling prediction of relativity can be illustrated by the following science fiction anecdote:

A twenty-year-old astronaut takes a trip to a faraway star in a spacecraft that can fly close to the speed of light. His twin brother remains on the earth. After fifty years on earth, the earthbound twin goes to the spaceport to receive his astronaut brother. Both brothers are amazed and startled. The earthbound brother has aged fifty years. He now has gray hair and wrinkled skin. But the astronaut brother has aged only one year. The clock, the calendar, and the biological aging process on the spacecraft slowed down to one-fiftieth of its normal speed. The twins agree that the adventurous space traveler is now forty-nine years younger than his earthbound twin is![22] According to late Carl Sagan, an eminent astrophysicist, we humans do not experience time dilation in our everyday life. However, nuclear particles experience it when they travel close to the speed of light. Time dilation is measured by their built-in-clocks called decay time and science has validated it with experimental data.[23]

The US Naval Observatory in Washington D.C. is responsible for maintaining our time standard. In 1972 a physicist tested the concept of time dilation by carrying along four cesium-beam atomic clocks during his round-the-world trip on a scheduled airline. These clocks could be trusted to a few billionths of a second over the time span of such a voyage. When compared with matching ones that remained in the observatory, the clocks lost time. The lost time agreed exactly with Einstein's predictions.[24]

Until a few years ago, scientists thought that protons and electrons were indivisible "elementary particles," but when the particles collided at high speeds, scientists discovered that the particles were made up of yet smaller units. In particle-physics laboratories, the time interval for short-living particles like muons to decay into electrons and nutrinos can be recorded. Experiments have shown that the faster muons take longer to die than the slow-moving ones. This is precisely what the special theory of relativity states.

Based on such experiments, the twin paradox will not be a paradox at all if and when we are able to travel close to the speed of light. However, the theory of relativity also suggests that nothing can travel faster than light.

According to the general theory of relativity, in a field of very high gravity, time will grind to a halt. Such high gravity exists in black holes and their peripheries. What is a black hole? When a

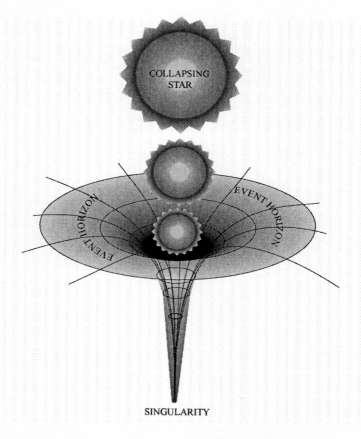

Figure 2-1. BLACK HOLE AND SINGULARITY. As the star implodes it increases in size. The matter within it is compressed into a region of zero volume, which then becomes a point where nothing exists. Here the curvature of space-time becomes infinite. At this point there is no time (no-when) and no space (no-where). At this point (called Singularity) all theories break down. The black hole, therefore, consists of an unobserved singularity within the event horizon of a star.

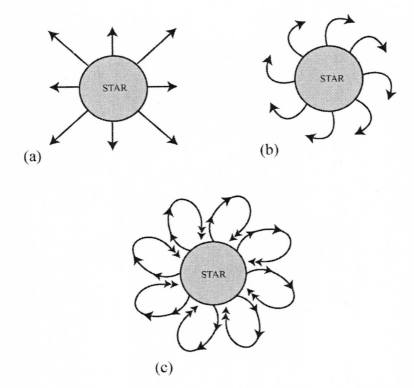

(a)

(b)

(c)

Figure 2-2. THE EFFECT OF GRAVITY ON LIGHT RAYS FROM A COLLAPSING STAR. As the intensity of gravity outside the collapsing star increases, the implosion of the star accelerates. The light rays leaving the star bend toward it. View A depicts the star before its collapse. View B depicts the bending of light rays toward the collapsing star. View C depicts the event horizon encompassing all the rays of light bending toward the star. For an observer located outside the event horizon, the star looks like a black hole because its light rays can no longer reach the sight of the observer.

dying star implodes or shrinks into itself under the relentless and unforgiving influence of gravity, the star becomes smaller and smaller. Indeed, the entire star is crushed out of existence at a single point. Physicists refer to the point of extinction as Singularity (Figure 2-1). At this point, the density of matter becomes infinite, and space-time is reduced to a mathematical point. During the shrinking process, the gravity around the imploding star becomes so strong that even the rays of light from the star cannot escape. Because we can see a star only when its rays reach our eyes, the imploding star literally disappears from our vision as it vanishes from the universe. What is left is called a black hole (Figure 2-2).

The periphery of the black hole is called the event horizon. We have no way of knowing what occurs inside the event horizon.[25] Theoretically time would be frozen for a person in an extremely high gravitational field such as the event horizon, while events continue to unfold in time for the observer on the earth.

The gravitational freezing of time could have interesting effects. Suppose, for example, that one of the astronauts started to sing the US national anthem while he was entering the event horizon. The general theory of relativity predicts that the situation for observers on earth (and anywhere outside the event horizon) would appear as though the astronaut has been singing the first note of "The Star Spangled Banner" even after the passage of billions of years. This prediction is flawed, however, because the time dilation would have slowed the sound vibrations to a degree whereby observers outside the event horizon would no longer be able to hear the note.[26] While time is frozen for the astronaut, events would continue on earth and outside it. The astronaut would not be able to leave the event horizon because, from his perspective, history has already advanced outside his space and time. He would be, literally, coming out before he went in. Similarly, according to the theory, two spatially separated events may appear to be happening differently for observers traveling at different speeds. One person may say Event A happened before Event B, and another would say Event B happened before Event A, while a third person might argue that Events A and B happened simultaneously.

When the Qur'an was revealed fourteen hundred years ago, people did not measure time as we do today. They did not wear watches that divided the day into time units, because clocks had not yet been invented. Their lives were not regulated by hours and minutes in abstract time but by the natural rhythms of nature, such as the changes in the seasons and position of the sun, even though water clocks, sundials, other devices were available to them. A day was the shortest span of time that had meaning and importance in the lives of the early Muslims, although the length of a day varied. Work began at sunrise and ended at sunset. Before the invention of the modern clock, they divided the day into five variable periods to mark the time for mandatory prayers based upon the position of the sun.

The study of the Qur'an guides us to understand the concept of time among early Muslims. Such a study is essential because time is inseparable from human activity and thoughts. Muslims believe that God is timeless. For example, Muhammad Asad, a contemporary Muslim scholar, states clearly: "What men conceive as time has no meaning with respect to God because He is timeless, without beginning and without end, so that 'in relation to Him a day and a thousand years are alike'."[27]

This Muslim belief is based upon the verses: "This is God, your Lord; there is no god but He, the creator of all things. So pay homage to Him for He takes care of everything." (Qur'an 6:102)[28] "He begets not, nor has He been begotten." (Qur'an 112:3)[29] "Originator of the heavens and the earth . . . ." (Qur'an 2:117)[30]

Time as a separately identifiable entity within the framework of the universe came into existence with the beginning of the universe. If God is the only entity that has no beginning, God exists without a clock that divides time into past, present, and future. Muslims posit that God sees, hears, and knows all things in a perpetual Now. This conclusion is supported by the Qur'anic verse: "Verily a Day in the sight of thy Lord is like a thousand years of your Reckoning." (Qur'an 22: 47)[31] Therefore, the Qur'anic description of time and the Muslim concept of time are that time is relative.

The human understanding of time in the Hereafter Universe (Al-Akhirah) also suggests a relative nature of time. For example, the following Qur'anic verses depict the paradox of the twins to which we referred earlier. The astronaut twin is comparable to a person who is resurrected in the Hereafter Universe (Al-Akhirah). He feels that his few years of life and the billions of years of lifeless existence between his death and his resurrection have shrunk to a brief moment.

> The day these people see seems to them as though they had stayed (in the world or in the state of death) only for an afternoon of a day or its forenoon. (Qur'an 79:46)[32]

> On the Day when He shall gather them (unto Himself, it will seem to them) as if they had not tarried [on earth] longer than an hour of a day, knowing one another; [and] lost indeed will be they who (in their lifetime) considered it a lie that they were destined to meet God, and [thus] failed to find the right way. (Qur'an 10:45)[33]

> On a Day when He will call you, and you will answer by praising Him, thinking all the while that you tarried (on earth) but a little while. (Qur'an 17:52)[34]

When human beings are resurrected, they will be asked about the length of their lives on earth, and they will reply that they lived on it for a day or a part of a day. Furthermore, they will attempt to sidestep the question by saying, "ask those who are capable of counting time." This evasion is an indication of the dissolution of humankind's earthbound concept of time, upon resurrection. The following verse suggests such dissolution:

> Then Allah will inquire from them, "for how many years did you live on earth?" They will say, "We stayed there for a day or part of the day; but ask those who (are able to) count (time)." (Qur'an 23:112-113)[35]

According to the verses above, the resurrected human being, like the astronaut twin, will feel that the whole world—its centuries, its ages, its epics, and its events—has shrunk to but a moment, the length of which is known to God only. Human beings will exist in a timelessness that words cannot describe. How could one express time in a place where there is no space-time relationship? Discussing timelessness is like defining time as an entity that flows. If it moves, it must have speed that can be measured. How can we measure it? If we can measure it, how do we express it? Speed, when measured, is expressed as "miles/hour." To express the speed of time, we would have to write "hours/hour," which is, of course, meaningless.

We have seen earlier that according to the theory of relativity, each observer lives within his own personal time frame, which does not generally coincide with anybody else's. Therefore, the Qur'anic verses do not violate God's (or Einstein's) law of relativity and the physical fact of time dilation. In summary, Qur'an teaches that time is relative based upon the state of observer and who observes it.

Why did God create time with the material universe? Why did He make time relative? In the construction of a rational universe for humankind, God knew the logical problems worldly existence would cause for His intelligent creature, the human being. Perhaps the major logical problem is the appearance of a contradiction between attributes such as God's Omnipotence and Omniscience, and His Will to create humans with the faculty to discriminate and choose either goodness or evil. For example, if God chooses to alter the future course of an event, do not His omniscience and omnipotence stand in the way of human free will?

Omniscience is a characteristic of God according to many verses in the Qur'an. For example, chapter 2:255 states:

> God, there is no God but He, the Living, the Eternal Sustainer. Neither slumber takes Him, nor sleep. His is what is in the heavens and what is in the earth. He knows what lies before them and what is behind them. And they grasp naught His knowledge, but of what He wills. His Throne embraces the heavens and the earth and it tires Him not to uphold them both. He is the Sublime, the Grand.[36]

The verses like this and many more suggest that God knows all things at all times. Hence, some atheists and logicians may argue that God's knowledge, being absolute, cannot go wrong. He knew what Timothy McVeigh was going to do even before he was created. In short, they posit that God "programmed" Timothy McVeigh to kill innocent Americans in Oklahoma City. They argue that McVeigh and other evildoers cannot be considered guilty of any sin because they are only preprogrammed machines.

Because All-knowing God anticipated these logical problems for humans, He created time as a factor in our universe. He made time relative for all earthly creatures, including humankind; but all events in all time frames, from the beginning of the Creation to the end of the current universe are current events as far as Allah is concerned.

Jewish physicist-theologian Gerald L. Schroeder clarifies the above scenario using the true story of a star that exploded, producing a sudden burst of light 170,000 light years from the earth.[37] Astronomers call such a sudden burst of light from an exploding star a supernova. The world came to know about the supernova when a Chilean astronomical laboratory recorded the arrival of the light on earth after 170,000 light years of travel.

Between the time of the supernova and the arrival of the light from it, a multitude of events occurred. Neanderthals came and went; we, modern Homo Sapiens, took control of the earth; Indus valley, Maya, Greek, Roman, and other civilizations peaked and vanished; the Qur'an was revealed; Muslims developed the scientific method and trained European scientists, ending the Dark Ages and engendering the Renaissance; the Muslim empire crumbled; and the United States of America took the leadership of the world—all before the arrival of the light from the supernova on earth.

Schroeder describes an imaginary consciousness without mass that travels with the light from the supernova. Let us assume also that our superconsciousness carries an internal clock with no mass as its companion. How much time would this superconsciousness have experienced? How many ticks would the superconsciousness's clock have made? Schroeder answers: "No time would have passed. Not a few years, not a few hours, or a few seconds. Zero time."[38] In this scenario there is no lapse of time for the superconsciousness, in spite of 170,000 years of recorded history.

A similar situation of Perpetual Now or Present exists in an environment with extremely high gravity, as in the case of a black hole. The 170,000 years of the earth, its epics, individual events, Timothy McVeigh and his atrocious bombing of the Alfred P. Murrah Federal Building as well as your and my life histories are experienced simultaneously by the superconsciousness.

The superconsciousness would definitely reject our claim that there is a history as well as a past and future series of events on earth. He would say that he sees everything that the primitive twenty-first-century human described as the past and the future. All happened in no time in his frame of reference. There is no post- or pre-knowledge for the superconsciousness, only omniscience, because all knowledge occurred at the same time. Physicist Gerald L. Schroeder freely admits, "I don't pretend to understand how tomorrow and next year can exist simultaneously with today and yesterday. But at the speed of the light they actually rigorously do. Time does not pass."[39]

If gravity, speed, black holes, and other material factors can affect and freeze time, is it conceivable for human beings to perceive the timeless world of God, who created the universe and time from nothing with His Enormous Infinite Power? A reasonable person would not deny the absence of time in the presence of God, the source of all forces of the material world. Accordingly, Islamic theology describes a perpetual Now that pervades in the spiritual domain of God, a state of being/consciousness that is more accurate than the "now" of our superconsciousness. Allah experiences all events of the material world in a perpetual Now without the linear flow of time that pervades the material world. God eternally knows all events as well as all their possible directions, deviations, and modifications occurring in the time frame of the material world. Until we fully understand physics's mathematical mystery of a concurrent yesterday, today, tomorrow, and next year, existing together at the speed of light as described by Schroeder, the coexistence of human free will and omniscience of God will remain incomprehensible because humankind has no experience of timeless existence.

According to the theory of relativity, nothing can travel faster than light; the theory negates any physical communication with a speed faster than light. This limitation in communication between two points is irrelevant from the Qur'anic point of view. The Qur'an states: "To God belong the East and the West. Wherever you turn the glory [face] of God is everywhere: All-pervading is He and all-knowing (Qur'an 2:115)."[40] God, being "All-pervading," is present at all points and at all events. He is present in every frame of reference, whether on the earth, in the light from a supernova, in black holes, or in the time frame of the astronaut of the twin paradox. God extends His presence even to our subconscious mind, as made clear in the verse "Whether you loudly avow the utterance (or not), surely He knows the secret and what is concealed behind that. (Qur'an 20:7)[41]

God's pervading presence in all frames of reference implies instant communication at any particular point we can imagine. When God comes though a particular frame of reference—for example, the earth—human beings picture that He is communicating in our local time. As we have seen in the previous chapter, Allah allows the next tick of the local clock to occur in order to open the door to the future with possibilities for His creations. Each tick of the clock is a knock at the door of the material world by messenger-moments of the future arriving with Allah's proposals. His creatures actualize into their timeline these possibilities arriving from the timeless realm of God. Hence the free choice of you and me and Timothy McVeigh.

An analogy from the science of molecular biology may better illustrate this concept. Chromosomes, with their component parts called genes, are the seat of all knowledge or information that creates the outward appearance of all living things as well as all chemical processes that take place within them. The genes are made of DNA. The genetic information in the gene, or DNA, has to be expressed in enzymes (a particular class of proteins) that guide all chemical processes in the cells. Intermediate molecules, called messenger RNA (mRNA), convey the information held by genes to these proteins.

A chromosome's particular knowledge or information is expressed though a chain of events. A gene for specific information is copied into a messenger RNA and sent to a ribosome. The ribosome manufactures (synthesizes) the protein to express the genetic information (or in metaphysical terms, the will) into body parts during fetal development and into the metabolism in an animal or human body.

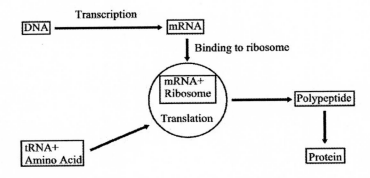

Figure 2-3. GENE EXPRESSION AND PRODUCTION
OF PROTEIN

The knowledge or information within the messenger RNA is a duplicated copy of the knowledge of a particular gene. Messenger RNA does not create new information or knowledge, but it acts as carrier of the information. Similarly, God, who knows all possible alternatives available in any given situation, presents copies of His knowledge to His creatures as possibilities packed in each moment of the arriving future from Him.

The ribosome uses the messenger RNA, a copy of a particular gene, to create proteins, body parts, and external characteristics of all organisms. Human and other creatures also use information presented to them by the divine activity in His timeless realm. Creatures actualize a particular possibility from the many possibilities coming as information within each moment of the arriving future, into the monuments of divine creation in the time and space of material world. Each moment of the arriving future is a messenger, like the messenger RNA of molecular biology. While one messenger RNA carries only one possibility, each messenger-moment of the arriving future carries more than one possibility. Each moment can carry contrasting as well as diametrically opposite possibilities or information.

We have seen that Muslims believe that the future never arrives until God creates it. The determining factor in the integration of the messenger RNA into a ribosome is the ribosome's prerogative, which science describes as a chemical property. In much the same way, the determining factor for actualization of any one of the possibilities into a given space and time is the free will that is a divine generosity to God's creatures.

God has the full freedom to determine the general direction and trend of the universe because He creates the messenger-moments of the arriving future, and He alone determines the nature of the possibilities to be packed in the messenger-moments for His creatures to actualize into monuments of God's creation. However, omnipotent God voluntarily and gracefully limits His absolute power and omniscience to create free will for His creatures and to objectify their choice of information in the arriving messenger moments from Allah.

To summarize, two contradictory realities exist:

1. The past, the present, and the future are real in the world of humankind;
2. No past, present, and future exist in the spiritual domain. There is only a perpetual Now in the world of God.

In the world of God, the Big Bang to the Day of Judgment and beyond are current events, and such human linguistic phrases as "what is intended to do" or "going to do" do not exist. Such a state of perpetual Now has no future or past. Einstein's own words are more revealing: "[T]he distinction between past, present, and future is only an illusion, however persistent."[42]

Jalaluddin Rumi, a well-respected Muslim Sufi, stated centuries before Einstein in his *Masnavi:*

> In the space-less realm of light of God, the past, present, and future do not exist. Past and future are two things only in relation to you; in reality they are one. Thy thought is about the past, and future; when it gets rid of these two, the difficulty will be solved.[43]

The poet Rumi rejected the idea of the labeling of time into past, present, and future and described it as only an illusion. Persians describe Rumi's thirteenth-century *Masnavi* as "the Koran in Persian."[44] He may have based his conclusions on the Qur'anic verses quoted above or on the verse "He knoweth what (appeareth to His creatures as) before or after or behind them" (Qur'an 2:255)[45], which implies that our past, present, and future are embodied in God's Now. The verse also suggests that God can see, in a moment, numerous events that are spatially separate, although for human beings the same events seem to occur one after another over a long period of time. God sees or knows the past, the present, and the future of humankind as a human being would see what is passing before his eyes at a given moment of "Now." There is no pre—or post-knowledge.

The Qur'anic verses, Rumi's *Masnavi,* and the theory of relativity inform Muslims that the twenty-billion-year-old universe could be seen as a six-day-old universe as indicated in the Qur'an. The "biological evolution of life over a period of three and a half billion years" could be experienced in "a blink of an eye" in a different time frame or in no time. This is further theologically supported by the following verses of the Qur'an:

> Indeed everything We have created in measure. And Our behest is one, as the wink of an eye. (Qur'an 54:49-50)[46]

Muslim scholars enjoy the fruits of scientific research—atomic energy, electronic innovations, medical breakthroughs, and so on. They accept that modern conveniences came out of the fundamentals of physics, chemistry, and biology. Muslim scholars also accept current modes of transportation, such as cars, trains, and airplanes. Likewise, we must also integrate the law of relativity with Muslim theology and recognize that such integration does not contradict the Qur'an, but is, in fact, endorsed by it. Once Muslims integrate the law of relativity with their beliefs about Creation, the contradictions between Muslim theology and modern knowledge evaporate completely.

CHAPTER 3

Creation and the Fate of the Universe

Cosmology, in the broadest sense, is that branch of learning that conceives of the universe as an ordered system. The term is derived from two Greek words, *kosmos* (order, harmony, the world) and *logos* (word, discourse). For a long time cosmologists thought that the earth was the center of the universe and that the Sun, the Moon, and the planets revolved around Earth.

In the sixth century the Greek Milesean school of pre-Socratic philosophers, which included Thales, Anaximander, and Aneximenes, taught that the formation of the world occurred as a natural, rather than a supernatural, sequence of events. They were followed by the Atomist school led by scholars such as Leucippus and Democritus, who described a boundless universe in which the interplay of atoms brings into existence endless worlds in various stages of development and decay. The Atomists did not admit the existence of a final cause.[1]

Aristotle viewed the universe as eternal, everlastingly undergoing change. He attributed changes on the earth to the motion of the heavenly bodies. Aristotle believed in a prime mover of the universe that keeps everything in motion and in constant change. He saw this being as nonmaterial and without potentiality. Further, he saw the mover of the eternal universe as a perfect entity.[2]

I recognize the importance of the Judeo-Christian interpretation of the Creation because, at present, creation-evolution arguments are often centered on the Biblical account of the genesis of the universe. I freely admit, however, that my knowledge of the Holy

Bible is limited and that many may find the following description of the Judeo-Christian view insufficient, considering the number of references in this work.

Until the middle of the nineteenth century, the relationship between science and religion in the West was "largely harmonious, and science took a subordinate role." Western scientists believed that their rightful task was to reinforce the central doctrines of Protestant Christianity.[3] Therefore, the Western cosmology of the nineteenth century was based largely on the Book of Genesis.

In chapter 1, verses 1-31 of the Book of Genesis, we are told that God created the universe in chronological order as follows: Initially the universe was without form. On the first day God created light and separated it from darkness. On the second day He created the skies and the oceans. On the third day He divided the earth into dry land and water, and then He created vegetation on the earth. On the fourth day God created the Moon, the Sun, and the stars. On the fifth day He created aquatic animals and birds. On the six day God created earth creatures such as reptiles, cattle, and wild animals, and He also created Adam and Eve. Finally God rested on the seventh day.

The order of events for the creation of the universe is not consistent. Genesis 2 tells us the universe was created in one day. chapter 1 says God created the fruit trees, the animals, Adam, and then Eve. Chapter 2 says God created Adam, the trees, the animals, and then Eve.

Modern scientific theories do not support the Biblical cosmology of the Book of Genesis, which holds that the Sun and other stars came into existence after the formation of the earth. In contrast to science, the Bible also contends that vegetation and water appeared on Earth before the formation of the sun and other stars, and that birds existed before reptiles. As a result of these conflicting claims, new interpretations of the Book of Genesis started to appear in the West.

Since Christians and Jews believe that God created the universe, they are considered creationists. However, they have no monolithic agreement among them about the different interpretations of Biblical verses. Based on their interpretations of and attitudes toward scientific theories, Christian creationists fall into four major groups: revelatory theorists, poetic theorists, old-earth theorists, and young-earth theorists.

Those who follow the revelatory theory propose that Genesis 1:11 is not a straight narrative account but that of Moses' vision of the divine revelation (Moses being traditionally believed to the human author of the Book of Genesis).[4] The second group of creationists follows the poetic theory. They claim that the story of Creation is a deliberate poetic account rendered by Moses and therefore should not be understood literally.[5]

The third group is the old-earth creationists, who attempt to justify a move away from literal interpretation. They see humanity as God's special and direct creation. They are distinguished from mainstream Christian accommodation to scientific consensus by their rejection of all or part of the theory of evolution. In general they accept the idea of an antique earth, based on a radiometric method, and have different explanations for the age of the earth and universe.

The old-earth creationist groups can be divided into three subgroups. The first, those who follow the gap theory, say that verse 1 of Genesis ("In the beginning God created the heaven and earth") suggests that God created an ordered universe. On the other hand, verse 2 ("And the earth was without form, and void . . .") reveals the act of giving form and order to an already existing chaotic universe. They believe that in the gap between what was revealed in verse 1 and verse 2, many cycles of creation and destruction occurred over millions of years, which allowed the formation of fossils from the creatures of the destroyed pre-Adamic world.

The second subgroup, who follow the day-age theory, take a metaphorical approach, proposing that the English translation of the Hebrew word *yom* as "day" is misleading. They believe that "yom" in Genesis does not constitute a twenty-four-hour earthly day, but a longer period of time. However, they reject the evolution of life and human beings.

The third subgroup follows the progressive theory and moves further toward accepting the scientific consensus. They see the days of creation as overlapping and of varying length. They concede that the order of events for the process of creation as revealed in Genesis differs from the order apparent in the fossil and geological records. In Genesis, for instance, flowering plants are created before land animals, but they are relative latecomers in the fossil record. Another Biblical inconsistency that progressive creationists accept is the creation of plants on the day before the Sun. Some interpret Noah's flood as a regional rather than a worldwide deluge. Nevertheless, for them, accepted scientific theory cannot explain major events such as the appearance of mammals or other groups. Instead these events require direct divine intervention. It follows then, that creation did not happen all at once but over time and that God gradually created new forms of life that eventually culminated in the creation of humans. Some individuals and groups espouse various combinations or variations of the three subgroups. Jehovah's Witnesses, for example, accept both the day-age and gap theories.[6]

The fourth, and dominant, group propounds the young-earth theory. They follow a highly literal and straightforward reading of the first eleven chapters of Genesis. They believe that the creation of the universe and life herein occurred in a span of six twenty-four-hour earthly days. Some of them believe that creation took place as recently as Archbishop Ussher's traditional date of 4004 B.C. James Ussher was an Anglo-Irish bishop (d.1656) known for his work on the chronology of the Old Testament. Other young-earth creationists acknowledge an age of ten thousand or even twenty thousand years for the universe. They believe that God created the universe in its present form and shape, and with all things in it, from nothing. Strict creationists assertively reject any large-scale evolution such as that of humans evolving from apes. They explain the geological record of sedimentary rocks and fossils as remnants of the creatures destroyed during Noah's flood.[7]

Additionally, there are those who believe that God created the universe but relinquished His spiritual authority over the natural and physical science. Finally, there are those who take the position that religion and scientific data are two different kinds of knowledge. Evolutionary theory and modern cosmology, they say, are not moral or religious doctrines, and the Bible is not a science textbook but a source of moral and spiritual truth. They assert that Genesis does not give a scientific account of the origin of the universe but rather conveys how God's creative act began and sustains it.

The Christian theologian Marcus Dods expresses such views in *An Exposition of the Bible*. Dods distinguishes between knowledge that is interested in the physical formation of the earth and that which is concerned with the relationship between the world and God. The first can be studied through scientific means, the second solely through Scripture (the Bible). He points out that although the details of the story of Creation in Genesis are irreconcilable with scientific discovery, they are "harmonious in their leading ideas." He contends that the compiler of Genesis was less concerned with the process of creation or scientific accuracy and more interested in conveying "certain ideas regarding man's spiritual history and his connection with God."

Dods asserts the futility of trying to reconcile science and religion. He points out that the twofold character of the story of Creation presented in Genesis illustrates two points: that the Bible is irreconcilable with scientific discovery and that it is in conflict with any pre-scientific cosmogonies. He adds: "Science knows nothing about the creation of the sun, the moon, and stars subsequent to the creation of the Earth. Likewise, science rejects the existence of fruit-bearing trees prior to the existence of the sun." Dods confirms that the writer of the Book of Genesis uses his knowledge for the purpose of conveying "his faith in the unity, love and wisdom of God the Creator." He states, "It is mischievous to build the teaching of the Book of Genesis around discoveries of science because it would not convince independent inquirers."[8]

BIG BANG THEORY

In the past, people believed that the stars were merely luminous objects affixed to the solid sphere of the sky. In 1718, when Edmund Halley noted that three of the brightest stars had changed their positions relative to the remaining stars, Western scientists realized that stars are moving objects.[9] They began to explore the universe and its origin but even today agree that our knowledge of the origin and evolution of the universe as a whole and its subunits is still very limited.[10]

In 1927 George Lemaitre claimed that "the universe has evolved from a hot dense state (about 10,000 billion degrees) and that the evolution commenced with a primordial explosion known as the Big Bang."[11] While the reason for the explosion remains still a mystery, it is clear that all matter and energy which constitute our universe today was compressed into a fiery mass of unimaginable density to form a cosmic embryo.[12]

Physicists and astronomers believe that insight into the structure of the atom will provide the ultimate answer to inquiries regarding the nature of the cosmic egg and the origin of the universe. Atoms have a longer history than that of the earth because they are the elements from which the earth is made. Before atoms came together to form the sun and its planets, they were drifting in the interstellar clouds. Therefore we can come to understand the genesis of the universe by probing into the constitution of atomic and subatomic particles, and the more we learn about the past of our universe, the closer we get to discovering the secret of its origin. Because it takes thousands or billions of years for light to reach us, events or stars that we see in the sky are hundreds or billions of years old. These stars and galaxies are fossils that guide us to understand the creation and evolution of the universe. Even common objects such as a pebble or a leaf provide us with clues because they, too, are made out of galaxies of atoms.

The structure of the atom can be pictured as a tiny solar system in which the sun is the nucleus and the planets are negatively

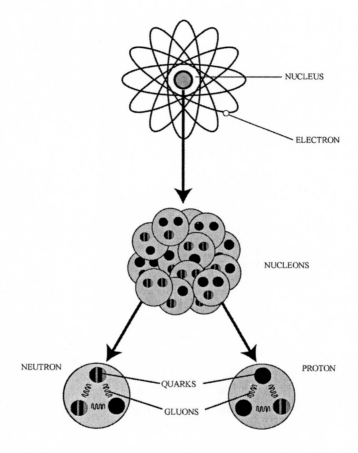

Figure 3-1. STRUCTURE OF ATOMS. Atoms, which constitute all matter, are made of nuclei and electrons. Nuclei are made of protons and neutrons, collectively called *nucleons*. Nucleons are made of quarks. A positively charged proton is made of two up-quarks and a down-quark; the neutron, a neutral particle, is made of an up-quark and two down-quarks. Gluons tie quarks together.

charged electrons, each rotating in a specified orbit (Figure 3-1). The nucleus lies within the confines of the orbiting electrons, and it is assembled out of neutral particles (*neutrons*) and positively charged particles (*protons*). When an electron jumps into another orbit, the photon carries a quantum of electromagnetic force between the orbits.

The elementary particles of an atom are made out of older particles known as *quarks*. We can learn more about subatomic particles by accelerating and colliding them in an electromagnetic field, which creates a wide variety of these particles.[13]

Subatomic particles interact with the aid of four fundamental forces: gravitational force (gravity), electromagnetism, a strong force, and a weak force. Physicists believe that these four forces, in conjunction with atomic particles, are the screenwriters of every event in the universe.[14 & 15]

Gravitational force keeps the physical components of the universe together by mutual attraction. A gravitational particle known as a *graviton* carries the force between two material particles. Since the graviton has no mass of its own, the force it carries is long-range. The gravitational force between stars and planets is attributed to the exchange of gravitons between the particles that make up these bodies.

Particles called *photons* carry the electromagnetic force by which electrically charged particles interact. Electric charges are of two kinds—positive and negative. The force between two charges is attractive only when one is positive and the other is negative. The electromagnetic attraction between negatively charged electrons and positively charged protons in the nucleus causes the electrons to circle the nucleus of the atom.

When an electron in an atom jumps from a higher-energy outer orbit to one of lower-energy that is close to the nucleus, the energy difference escapes in the form of a photon, which we see as light. The atom cannot exist without the strong nuclear force, which bonds the quarks to form protons and neutrons and holds protons and neutrons together in the nucleus. This force is carried by another particle called a *gluon*, which interacts only with other gluons or with quarks. Finally, the weak nuclear force, carried by a particle known as *boson*, helps the neutrons inside the nucleus to decay. The strong and weak forces operate only in the nuclei of atoms, but the range for electromagnetism and gravitation is infinite.[16]

Physicists believe that when the Big Bang occurred, all four forces—gravitational, electromagnetic, a strong force, and a weak force—united under extremely high temperatures. This process remains a mystery because no laboratory on earth can produce such temperatures. Only the high ambient temperatures of the genesis were hot enough to create such unity. The universe then started to expand and cool. Gravitation became a distinct force 10^{-43} seconds later. Further cooling to $10^{28°}$ Kelvin, at 10^{-35} seconds after the Big Bang, the strong force became distinct. After about 10^{-10} seconds, electromagnetism and the weak force emerged. (Figure 3-2). When the universe was one second old, quarks were formed, and when it became one hundred seconds old, the protons and neutrons bonded to form hydrogen and helium. One billion years after the Big Bang, the galaxies formed.[17] (Figure 3-3).

Physics does not have any explanation for the beginning of creation, when everything in the universe was condensed to a singularity in which the density of matter became infinite and space-time was reduced to a mathematical point. In the singularity, the extreme heat unified all four forces. Similarly the origin of the four fundamental forces and quarks remains a mystery. Are the four fundamental forces really fundamental? Is the quark the final elemental particle that is indivisible?

After the colossal explosion called the Big Bang, the universe began an expansion that has never stopped. Edward Hubble, who was inspired by the red shift of the galaxies, discovered evidence for the expanding universe in 1932.

In 1842 Christian Doppler pointed out that the frequency of the sound we hear is higher when a train is approaching and lower when it is moving away. This is known as the Doppler

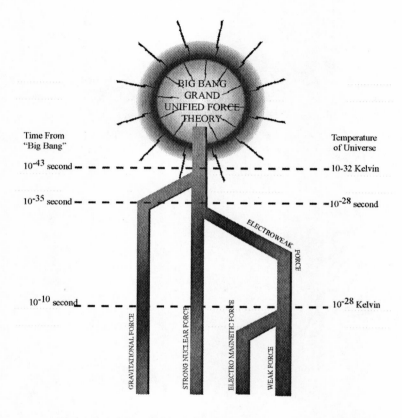

Figure 3-2. SYMMETRY BREAKING. A theory or atomic process is said to possess symmetry if it undergoes no change when certain operations are performed on it. Such a state existed when the Big Bang occurred. The four forces—gravitational, strong, weak, and electromagnetic—united into one force (Grand Unified Force or Theory). As the ambient temperature cooled following the Big Bang, the Grand Unified Force disassociated into four forces.

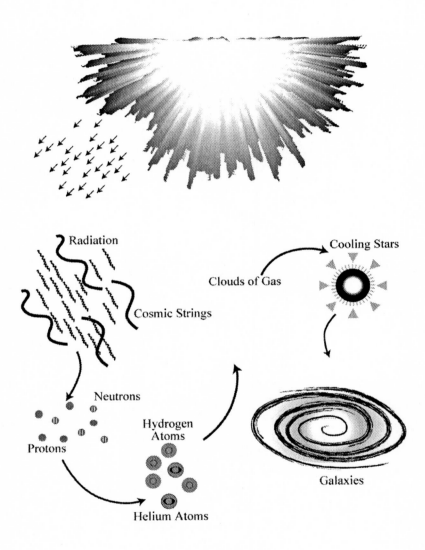

Figure 3-3. BIG BANG.

effect. The French physicist Armand Fizeau showed that the Doppler effect works for light as well. The light frequency "red" is longer when the source of light moves away from the observer and shorter when it approaches.

When light passes through a spectroscope, light waves spread out. The longest wave (red) will be at one end, and the shortest wave (violet) at the other end. Light waves between the red and violet waves are based on various frequencies. Therefore, we get the following order of color—red, orange, yellow, green, blue, and violet. Where some wavelengths are missing, we see dark

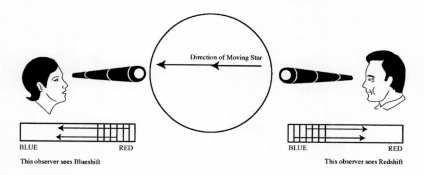

This observer sees Blueshift   This observer sees Redshift

Figure 3-4. DOPPLER EFFECT. Radiation from an approaching star is compressed to shorter wavelengths than usual, (〜〜〜) and the dark lines shift to a blue color (blue shift). Radiation from a receding star is stretched to longer wavelengths (〜〜〜〜) than usual, and the dark lines shift to a red color (red shift).

lines on the spectrum. When a star moves away from Earth, the dark lines shift toward the red end of the spectrum. This is called *red shift*. If the star gets closer to the observer, the dark lines shift to the violet end of the spectrum (Figure 3-4). The observation of this phenomenon led Hubble to the conclusion that different galaxies are moving away from each other and that the universe is expanding.[18]

THE QUR'AN AND THE BIG BANG

We have reviewed the antiquated history of cosmology that existed prior to that based on new physics. We also reviewed the Biblical cosmologies. The Qur'an, the Muslim's word of God, does not have a chapter on genesis but has verses that deal with Creation.

The Qur'anic concept of the universe differs from that claimed by Herman Bondi, Thomas Gold, and Fred Hoyle, who claim that ours is a steady-state universe that never had a beginning and will continue to exist in a condition similar to its present. The universe, according to the Qur'an, has an origin and an end, as we shall presently discuss.

Knowledge of the external world is reflected in the sophistication of the language that people use. We invent new words to convey new knowledge. When the Qur'an came to Arabia, the people of that time and place used "the heavens and the earth" to describe the universe. The Qur'an describes the heavens and the earth (the universe) as a single entity that God split into the heavens (galaxies) and the earth.

> Do not these unbelievers see that the heavens and the earth were an integrated mass, then We split them and made every living thing from the water? Will they not believe even then? (Qur'an 21:30)[19]

The Qur'an further implies that the universe evolved from a gaseous state that is described as "smoke."

> And He (it is who) applied His design to the skies, which were [yet but] smoke; and He [it is who] said to them and to the earth, "Come [into being], both of you, willingly or unwillingly!"—to which both responded, "We do come in obedience." (Qur'an 41:11)[20]

Science has come to agree with what the Qur'an has described in the sixth century. Most astrophysicists concur that the universe began with the Big Bang. Early universe was filled with radiation and elementary particles, such as quarks. Then, hydrogen and helium were formed from the elementary particles in the inferno of the Big Bang. They believe that the gas clouds condensed and gave birth to stars. These eventually took shape to form the galaxies.[21]

In spite of this convergence of the scientific perception of the origin of the universe and the literal Qur'anic concept of Creation, many Muslims of our time are afraid that the Big Bang theory could be replaced by another paradigm and that if that happened, their faith might be weakened. Some modern Islamic scholars such as Muhammad Asad argue, "It is, as a rule, futile to make an explanation of the Qur'an dependent on scientific findings, which may appear true today but may equally be disproved tomorrow by new findings."[22] Whether or not the Qur'anic verses agree with modern science, these verses describe an explosive beginning from an integrated mass or singularity.

The Qur'an also describes the universe as one that is expanding:

> And it is We who have built the universe with [our creative] power; and it is We who are steadily expanding it." (Qur'an 51:47)[23]

The Qur'an reveals many signs and events that would take place before the universe ends. The following verse states that the end is heralded when the *currently* "receding stars" and planets are commanded to reverse the direction of their courses: "So, I call the receding stars to witness, The planets withdrawing into themselves. (Qur'an 81:15-16)[24]

The verb خنس (khunasa), used in the Arabic text of this verse, means "to recede."[25] The Qur'anic statement of the reversal of the direction of the movement of the stars is evidence that the universe is currently expanding. Here, too, astrophysicists seem to agree with the Qur'an when they state that following the huge explosive beginning, the universe began an expansion that is still ongoing today.[26]

The majority of astronomers generally agree on how the universe began, and the evidence from the phenomenon of red shift, described above, is the basis for the scientific assumption that the universe is expanding. Their views differ, however, on the fate of the universe.

Astronomers have two theories on the subject. One holds that the expansion of the universe is not self-limited and will continue. The galaxies will fly away from each other, space will become emptier, and eventually some stars will run out of nuclear fuel. They will shrink, cool, and end up as black dwarfs. Some stars will die as supernovae, exploding and releasing tremendous energy. The debris of the explosions will crumble inward due to immense gravity, to form neutron stars. The heavier neutron stars will again be crushed and shrunk due to gravity until they finally become invisible black holes. Nobody knows how black holes die.

The other view rejects the above scenario and holds instead that the expansion of the universe will come to a halt at some point in the future. When this time comes, the outward momentum of the primordial explosion and the inward pull of gravity will momentarily hold the universe in a balance. Then gravity will pull the components of the universe together at an increasingly faster pace until the universe collapses into the heat and chaos from which it emerged a few billion years ago. [27]

Some astronomers believe that the universe will not come out of the collapsed state, while others speculate that it will rebound with a new explosion, experience a new moment of creation, and "become an entirely new world, in which no trace of the existing Universe remains."[28]

The Qur'an acknowledges an end to the universe; the fundamentalist Christian view of the fate of the universe does not. The Bible presents the destiny of creation as eternal. Psalms 148:3-6 state:

3. Praise ye him, sun and moon: praise him, all ye stars of light.
4. Praise him, ye heavens of heavens, ye waters that be above the heavens.

5. Let them praise the name of the LORD: for he commanded, and they were created.
6. He hath also stabilized them *forever and ever: he hath made decree which shall not pass.*

Gerald L Schroeder, an applied physicist and a theologian who received a doctoral degree from Massachusetts Institute of Technology, comments:

> Based on these verses, sages have said that the universe can exist forever. But if its destruction is to come, it will not be a natural act. It will be a special force from God as was its creation. The heavens are eternal. This is true according to the Bible's description of our universe and according to the best estimate that science has for the situation.[29]

The following excerpt from "Did Man Get Here by Evolution or by Creation?" a publication of Watch Tower Bible & Tract Society, suggests that our universe will last forever:

> With bad conditions and violently wicked men no longer existing, the whole earth will become the possession of people who love righteousness. The Bible states: 'The meek ones themselves will possess the earth, and they will indeed find their exquisite delight in the abundance of peace. *The righteous themselves will possess the earth, and they will reside forever upon it.*'—Psalm 37: 11, 29. That is how God will more than compensate for the wickedness that innocent people have suffered during their lifetime. Throughout eternity God will shower down blessings on mankind, so that any hurt they have received in the past will fade to dim memory, if they care to remember it at all. The Creator guarantees: 'I am creating new heavens (a new heavenly government) and a new earth (righteous

human society); and the former things will not be called
to mind, neither will they come up into the heart. But
exult, you people, and be joyful forever in what I am
creating.'—Isaiah 65: 17, 18.[30]

Thus, the literal interpretation of Jehovah's Witnesses regarding
the fate of the universe is that after the second coming of Jesus, the
wicked will be destroyed and the righteous will inherit and "reside
forever upon" the earth. If the righteous are going to stay forever
on the earth, the earth has to be eternal, too.

While the scientific data on the fate of the universe is derived from
piecemeal experiments and observations by modern scientists, the
Qur'anic description of the fate of the universe can be deduced from its
account of the events that will take place on the Last Day (*Yawm al-
Qiyamah*). It has been explained earlier (see chapter 1) that the Last Day
should not be conceived of as the earthly twenty-four-hour day but as
an aeon or a long period of time. The descriptions of the events on the
Last Day are scattered over many chapters and verses in the Qur'an.

The following verse indicates that the universe, including the
earth, has an "appointed term" for existence:

> We created not the heavens and the earth and all between
> them but for just ends, and for a term appointed . . .
> (Qur'an 46:3)[31]

The Qur'an describes the events of the Last Day, but it does
not explain how these events come about. We know, however,
that they follow the physical and chemical laws of the universe.
We have seen earlier that the Qur'an in chapter 29, verse 20, asks
the believers to *"journey in the land, then behold how God originated
creation."*[32] According to this verse, the objects and method of
creation are understandable. Therefore, the process and events at
the end of the universe are also understandable.

It is tempting to quote verses from the Qur'an on the events
of the Last Day (*Yawm al-Qiyamah*) because quotes from modern
scientists appear to repeat verbatim pertinent Qur'anic verses. One

scientist claims that "in some tens of times 10^9 years, the recession of the neighboring galactic systems will cease, and the process will start to reverse."[33] Another scientist contends that the universe will start contracting and that "at first the contraction will be slow, but over many billions of years, the pace will accelerate."[34] The following Qur'anic verse seems to introduce, in the common layman's language of the Arabs during the sixth century, the scientific facts of our age:

> On that day when We shall roll up heaven as a scroll for
> writings; As We originated the first creation, so We shall
> bring it back again—A promise binding on Us; So We
> shall do it. (Qur'an 21:104)[35]

The above verse describes the universe as a written scroll of God. The metaphor of the scroll seems to be a figurative explanation of the collapsing universe of science. If we release an open scroll, it will recoil slowly in the beginning and as it rolls back towards the end, the process accelerates.

After the Big Bang the galaxies began to recede. "The galaxies that are now receding from one another will start to approach each other gathering speed all the time."[36] Over fifteen centuries ago the Qur'an stated that the receding galaxies would reverse their courses on the Last Day. Fourteen centuries ago, the Qur'an stated in the chapter, Coiling Up [of the universe], that the receding galaxies would reverse their courses before the end of the universe: "So verily, I call to witness the stars that recede, (Qur'an 81:15-16)"[37] Likewise, as the result of the final "rolling up" of the universe, astronomers would see the stars advancing (blue shift) toward them rather than receding (red shift), as they see today.

During the final collapse, what happens to our solar system? Astronomers believe that under the combined influence of hydrogen fusion on the solar surface and the high temperature helium fusion in its interior, the exterior of the Sun will inflate and cool.[38] This cooling of the Sun may be what the Qur'an alludes to in a verse that describes the Last Day: "When the Sun shall be darkened" (Qur'an 81: 1).[39]

According to scientific interpretation, after the expansion and exterior cooling, the Sun will become a red giant star. When the Sun enlarges, it will swallow the planets Mercury and Venus—and probably the earth also. The inner solar system will then becomes a part of the Sun.[40] The following verse seems to describe the red bloated sun: "Upon the day when heaven shall be as molten copper" (Qur'an 70: 8).[41]

The Qur'an speaks of a merger between the moon and the sun during the last days of the universe: [On the Last Day] "And the sun and moon are united" (Qur'an 75:9).[42] It does not mention other planets of the solar system because the common Arabic language at the time of revelation did not include the words Mercury and Venus. However, all Arabs definitely knew the moon. Rotation of the moon was the basis for their calendar.

According science, billions of years from today when the collapse of the universe proceeds uninterrupted, the sun will gradually become red and inflated. Its overwhelming heat will melt the Artic and Antarctic ice resulting in coastal flooding throughout the world.[43]

Similarly, the Qur'an states that the last days of the universe will be heralded by rising water levels in the oceans and by flooding along the coasts: [On the Last Day] "And the ocean filled with swell" (Qur'an 52: 6).[44] The Arabic verb *sajara* in the text of this verse means, "to fill."[45] According to current cosmology, flooding of the continental coasts is the result of the "filling" of the ocean due to the melting ice of the poles.

As the collapse continues, the soaring ocean temperature will lift clouds of steam into the air. Finally, the unforgiving extreme heat from the sun will make the oceans boil.[46] The Qur'an seems to be ahead of its time in warning us that during the last days of the universe, the ocean will boil: [On the Last Day] "when the sea shall be set boiling" (Qur'an 81:6).[47]

Science tells us that during the collapse of the universe, the "observer would no longer be able to discern individual galaxies for these would now have begun to merge with each other as intergalactic space closes up."[48] This quote is similar to the Qur'anic revelation that the observable shining points in the sky will decrease during the merger of the galaxies such that the sky appears empty: [On the Last Day] "when the sky stripped bare" (Qur'an 81:11).[49]

Science tells us that as the end of the current universe approaches, initially the galaxies, then, the stars will collapse and merge into each other until it becomes a cataclysmic inferno. The continuing collapse of the universe will result in a singularity identical to where our universe originally began.[50]

The Qur'an, in the following verses, seems to narrate the events of the Last Day like a modern astronomer describing the events of the vanishing universe:

> The day the heaven shall be rent asunder with the clouds . . . (Qur'an 25:25)[51] (The "clouds" may be those originating from the boiling ocean.)

> [On the Last Day] When the sky cleft asunder, when the stars are scattered (Qur'an 82: 1-2)[52]

> [On the day] When the stars fall, losing their luster (Qur'an 81: 2)[53]

> And (on that day) the earth is moved, and its mountains, and they are crushed to powder at one stroke. (Qur'an 69:14)[54]

> When the sky is rent asunder and it becomes red like ointment (Qur'an 55:37)[55]

> It will be no more than a single blast . . . (Qur'an 36:53)[56]

The events that the Qur'an describes seem to fit the formation of black dwarfs, supernovae, and the final event of the universe collapsing into singularity.

What happens after the big crunch? When the collapse of the universe is complete, all the matter in the universe will be compressed into a single point. At this point of infinite space-time curvature, every atom and every particle in our universe is crushed out of existence.[57] This point becomes a state of singularity—a single point of space and time, where density and temperature become infinite and theories of science become invalid. One might speculate that there were new laws that worked at singularities, but, according to Stephen Hawking, it would be very problematic even to formulate such laws at such unruly points, and our observations would not help us to define what those laws might be.[58]

Whether the universe rises again phoenix-like from the singularity is not known.[59] Should the universe rise again, Carl Sagan asks: what would be the nature and properties of the new universe? Would the laws of physics and chemistry, that would govern the new universe, be different? In the universe that springs out of the Big Crunch, would there be stars and galaxies and world, or something quite different?[60]

One wonders whether the previously cited ayath describing the scroll answers the above questions:

> The day when We shall roll up the heaven as a scroll is rolled for writings; as We originated first creation, so *We shall bring it back again*—a promise binding on Us; So We shall do. (Qur'an 21:104)[61]

The literal meaning of the verse above suggests a new universe will spring out of the singularity. Similarly, the following verse could be the answer to Carl Sagan's questions.

> On the day when earth is changed into different earth and heavens into new heavens, Mankind shall stand before God, the One, who conquers all. (Qur'an 14:48)[62]

The verse suggests that, after the Big Crunch, the newly created universe will be based on new laws of physics and chemistry and maybe different mathematics. Two plus two may not become four. Also, the new universe of the Hereafter would not have time-space relationship as we have seen earlier.

In summary, we find much agreement between Qur'anic revelation and scientific discovery. Both maintain that the universe originated from nothing and evolved into its present state over a long period of an earthbound time. They agree that the universe is expanding and that at some remote future, it will collapse into a single entity or singularity where existing physical and chemical laws become nonexistent, mathematics would not add up, and numbers become infinite.

Not only the Qur'an but also some scientists believe in the rise of a new universe. Whether or not the new universe is created after this one collapses remains controversial. Scientists who grew up with Judeo-Christian traditions cannot rationally concur with the Biblical description of the origin of the universe, so they vehemently oppose the idea of the supernatural origin of the universe. For Muslims, however, the modern discoveries of science are less problematic.

CHAPTER 4

The Age of the Universe:

Uniformitarianism, Fossils, and Muslims

A prevalent publicly held assumption is that Jews and Muslims share the Christian fundamentalist view in the debate between science and creationism. Western scientists' lack of knowledge about Islam prompts them to preach scientific "truths" to all believers, including Muslims. For example, Maitland A. Edey, the editor-in-chief of Time-Life Books, and Donald C. Johanson, director of the Institute of Human Origins in Berkeley, California, proselytize:

> We talk today about 'Gospel truth' as the truest kind of truth, the kind we swear by, the kind we know in our bones to be true, the last truth of all that we are willing to give up. It was that kind of truth, a universal belief shared by all traditions, all prejudices, all faiths, that had to be unraveled and restated . . . . Indeed, if we truly believe in God, we should recognize that He gave us the brains to conduct scientific research. Not to do so—not to use the marvelous gift of intelligence with which He has endowed us—would seem to be disrespectful of God.[1]

If Muslims of our time followed what the Prophet Mohammed and the Qur'an command them to do, the above authors could have limited their admonishments to the Biblical literalists among Jews and Christians. The Prophet said:

> Seek knowledge; for its acquisition is fear of God, its pursuit is worship, its discussion is prayer, search for it is holy war, its propagation is charity, its teaching is fraternity. For it is the index of right and wrong; it is the lighthouse of the road to paradise. It provides consolation in loneliness, friendship in estrangement, fidelity through thick and thin, a protective arm against enemies, rapprochement towards foreigners, decor among friends [and] through knowledge, God is obeyed and worshipped. In knowledge the good is revealed, loved and implemented, for it is the prius of action, the first principle of every good deed.[2]

Few Western historians have acknowledged the above belief, which was an integral part of the faith of early Muslims. J. Bronowski states, "Muhammad firmly emphasized that Islam was not to be a religion of miracles. It became in intellectual content a pattern of contemplation and analysis."[3] This Islamic doctrine paved the way for the development of modern science and the Western Renaissance. The common Western belief that scientific thought originated in the West constitutes a misrepresentation of historical fact. Dr. Robert Briffault, a British historian acknowledges:

> Roger Bacon learned Arabic and Arabic science. Neither Roger Bacon nor his namesake has any title to be credited with having introduced the experimental method. Roger Bacon was no more than one of the apostles of Muslim science and method to Christian Europe, and he never wearied of declaring that knowledge of Arabic and Arabian science was for his contemporaries the only way to true knowledge.[4]

Accounts written by Western historians of science credit Lyell and Hutton for the theory of uniformitarianism, Leonardo da Vinci for the discovery of the origin of fossils, and Charles Darwin with recognizing the origins of humanity from the apes and the phenomenon of natural selection. To date Western historians have not given credit to brilliant Muslim scholars such as al-Biruni, Ibn-Arabi, Ibn-Khaldun, al-Jahiz, Ibn-Sina, Ikhwan al-Safa, Ibn Haithem. On the contrary, they claim that Judeo-Christian and Muslim scriptures teach that the Earth and universe as a whole are very young, and proclaim that Western scientists originated the theories enumerated above. Let us examine the validity of their claims.

Western historians maintain two views on the age of the universe: the Judeo-Christian and the scientific. Until the mid-nineteenth century, the West knew only what the Bible revealed. The Book of Genesis furnishes genealogical data in chapters 4, 11, 21, and 25. In addition, the Hebrew Bible gives durations for such variables as how long a particular king reigned, the age of a person when a son was born, and how many years elapsed from one event to another. Based upon this Biblical genealogy, Abraham was born 1948 years after the creation of Adam (Table 4-1).[5 & 6]

In addition to the above genealogy, some Biblical passages describe events to which other historians assign firm dates. This means that we can carefully work our way backward and perhaps calculate the year of the events with which the Old Testament begins.

One person who did this rather early on was an Irish-born Anglican bishop, James Ussher (1581-1656), who concluded that Adam and Eve were created by God in 4,004 B.C.[7] Others have improved on this, "pinpointing the actual moment of human creation at 9 A.M. on Sunday, October 23."[8] By adding the years between the creation of Adam and the birth of Jesus (4,004 years) to the years between the birth of Jesus and our time (2000), strict

Name	Date of Birth after creation of Adam	length of life in years	Date of death after creation of Adam	Source in Genesis
Adam	930	930		5:3, 4
Seth	130	912	1042	5:8
Enos	235	905	1140	5:11
Cain	325	910	1235	5:14
Mahalaleel	395	895	1290	5:17
Jared	460	962	1422	5:20
Enoch	622	365	987	5:23
Methuselah	687	969	1656	5:27
Lamech	874	777	1651	5:31
Noah	1056	950	2006	9:29
Shem	1556	600	2156	11:10,11
Arphaxad	1658	438	2096	11:14,15
Eber	1723	464	2187	11:16,17
Peleg	1757	239	1996	11:18,19
Reu	1787	239	2026	11:20,21
Serug	1819	230	2049	11:22,23
Nahor	1849	148	1997	11:24,25
Terah	1878	205	2083	11:32
Abraham	1948	175	223	25:7

Table 4-1. THE BIBLICAL GENEALOGY OF
ADAM TO ABRAHAM

creationists estimate that the universe was created roughly six thousand years ago. Some among them allow an age of ten thousand or even twenty thousand years for the universe.[9]

The pre-Renaissance Jewish scholarly calculation of the age of the earth is no different. The earliest portions of the Hebrew Bible were compiled when Jews were captives in Babylonia (586-539 B.C.). They incorporated the Babylonian version of primeval history, including the story of the worldwide Flood. By tracing the long lifespan of the antediluvian patriarchs and their sons, one might calculate the year of the creation of Adam and Eve. According to Jewish scholars, God created the universe in 3760 B.C.[10]

Just over a millennium ago, in 1000 A.D, Abu Raihan al-Biruni, a Muslim scientist, recorded the methods used by Jews and Christians to calculate the year of creation. One method was as described above. Al-Biruni described another method whereby Jewish and Christian scholars used the total numerical value of particular words in the Hebrew Torah or Syriac version of the Bible respectively, to calculate the years between Adam and Alexander the Great. He says: "The Jews and Christians differ widely on this subject; according to the doctrine of Jews, the time between Adam and Alexander is 3,448 years whilst, according to the Christian doctrine, it is 5,180 years."[11] We can calculate, therefore, by adding the years from Adam to Alexander, plus the years from Alexander to 2,000 A.D., that until recently the year of creation of the universe was 5,771 years ago for Jews and 7,506 years ago for Christians.

Many post-Renaissance Christians and Jews reject the calculation of the age of the earth based on Biblical genealogy. They claim that such calculations are not reliable because the genealogy is incomplete and documents only important people. They do not cite any references from antiquity to support their argument, however. If they have a reference, it is invariably from Jewish scholars who lived among Muslims.

Large numbers of Americans in our technological era still believe that the earth was created very recently. A late 1970s Gallup Poll found that 42 percent of the public believed that the Bible is literally true. Another national poll in 1982, reported in the *New York Times*, found that 44 percent of Americans agreed that God created man pretty much in his present form within the last 10,000 years. These polls indicate that creationist beliefs are not an extremist view, nearly half of the American population believes in them.[12]

When scientific knowledge about the ancientness of the earth, evolution, and uniformitarian principles (discussed below) started to spread in the Judeo-Christian West, new theological doctrines emerged to accommodate science. (These include the old-earth creationism based on the gap, day-age, poetic, and revelatory theories that we reviewed in the previous chapter.) Despite the rise of new interpretations, almost half of modern-day Americans continues to insist upon the recent origin of the universe, even though many contemporary Christian leaders, after taking into consideration the scientific evidence, argue for an accommodation of their faith with the modern science. Many Christian leaders, including Pope John Paul II, have rejected "the kind of strict literalism implicit in Ussher's approach is in no way a necessary concomitant of religious faith."[13]

One cannot subscribe to the idea of a six- or seven-thousand-year-old Earth for numerous incontrovertible reasons. For example, observance of a cross section of a tree shows an annual variation in its growth that is indicated by a series of concentric rings, one for each year of growth. The number of rings in the cross section is equal to the age of the tree. (Figure 4-1). The oldest known trees are the bristlecone pines. A study of these pines and their concentric rings shows that some of these trees are approximately eight thousand years old. If we accept Bishop Usshur's calculations, they are much older than the Earth.[14]

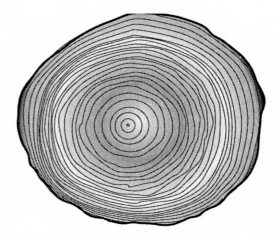

Figure 4-1. CROSS SECTION OF A TREE TRUNK. Note the annual rings of alternating light and dark wood. The xylem formed during each growing season forms distinct bands. They are called the *annual rings,* which can be counted to determine the age of the tree.

The speed of light is 186,000 miles per second. The distance that light travels in a year is called a *light year.* It takes eight minutes for light to reach Earth after it leaves the Sun. The *Voyager's* signals from Saturn reached Earth more than an hour and half after they were transmitted. The stars we see today are tens of thousands of light years away.

In fact, we see them as they were millions of years ago, when their light left them to travel toward us. Therefore, we must concede that these stars are several million years older than the six-thousand-year-old universe. The late George O. Abell, professor of Astronomy at the University of California, asks, "Unless, that is, we are to suppose that God created all those remote stars less than six thousand years ago and at the same time created light from them already well on its way to us. Why should God do this?"[15]

Another contradiction to the belief in a six-thousand-year-old universe is the continental drift, which is the large-scale horizontal displacement of continents, relative to each other.[16] At one time all the continents were a single landmass, but they gradually drifted from Africa (Figure 4-2). As scientist George O. Abell explains:

> We can now directly measure the motion of the continents. Laser satellite experiments show that North America and Europe are separating at about two centimeters per year [and] to separate the twenty-five-hundred-mile beach across the Atlantic at that rate (less than one inch per year) has taken approximately 200 million years.[17]

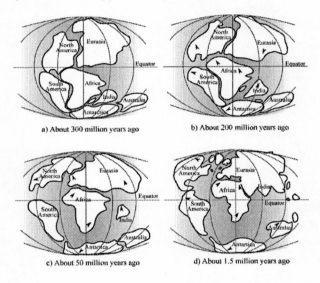

a) About 300 million years ago

b) About 200 million years ago

c) About 50 million years ago

d) About 1.5 million years ago

Figure 4-2. CONTINENTAL DRIFT. Movement of continents (indicated by arrow) produced the modern continents from a single landmass—Pangaea—that existed in the Paleozoic era (between 200 million to about 600 million years ago).

If we endorse the contemporary fundamentalist Christian and pre-Renaissance Jewish claim of a six-thousand-year-old universe as the truth, then we have to reject the scientific achievements of NASA and the physics upon which they were based. Scientific progress would halt abruptly, and the West would again fall back into the Dark Ages.

We have established that the universe is older than six thousand years. We have also seen the variable opinions among the Jews and Christians about the age of the universe. Contemporary Muslims, on the other hand, do not entertain possible interpretations of this issue. That is not surprising because since the beginning of the Muslim Dark Ages, they were not allowed to think for themselves. Only Muslim scholars (*ullamah*) were permitted to do so. Most Muslims believe that the ullamah are the only ones capable of interpreting the Qur'an and providing the world with the truth. Any opinion regarding spiritual or secular matters that does not correspond to that of the ullamah or individualistic attempt at interpreting the Qur'an is considered a heresy. Another reason Muslims are lacking in their knowledge of the age of the universe is the rampant ignorance and illiteracy among them.

The third reason is lack of freedom of expression in the contemporary Muslim world. Muslim rulers have bottled up the Muslim mind. Alas! The West now imagines and redefines that bottled-up mind in their books, newspapers, and magazines while Muslims at large remain in the bottle as mere spectators with no voice of their own. Those of us in America who have the freedom to reflect about the universe should ask: What is the age of the universe based on the literal reading of the Qur'an and based on Muslims who lived prior to the beginning of fourteenth century?

The Qur'an states that God created the universe in six days (*fi sithati ayyam*): "Your Guardian—Lord is Lord, who created the heavens and earth in six days . . ."(Qur'an 7:54)[18] The verse refers to a process that began before the earth and sun were created. It places the creation of the universe on a time scale. As discussed in chapter 2, the six days referred to above are not equal to six earthly days: "Verily a day in the sight of thy Lord is *like* a thousand years of your reckoning. (Qur'an 22:47)[19] The thousand years mentioned in the verse are not *equal* to a thousand years of our earthly time. The six days of creation means six long periods of earthly time. Even though we cannot precisely determine the age of the universe from the above Qur'anic verses, we can reasonably conclude that the universe was created a very long time ago. Does this inference agree with the opinion of early Muslim scientists?

Seven hundred years before James Hutton (1722-1799), the earliest Western author of uniformitarianism, Abu Raihan al-Biruni (973-1048) was born in Khiva, Uzbekistan. Al-Biruni was among the most educated and cultured men of his time, with a brilliant intellect "encyclopaedic in scope."[20] In addition to mastering the Arabic language in which he wrote, al-Biruni was proficient in Turkish, Persian, Sanskrit, Hebrew, and Syriac. He excelled in numerous fields of knowledge and particularly in astronomy, mathematics, chronology, physics, medicine, and history. He died in Ghazani, Afghanistan.

Al-Biruni believed that the age of the earth and universe was so great that he could not speculate on it. He criticized the then-held Judeo-Christian teachings of the six days of creation in Genesis as equal to the six earthly days of twenty-four hours each. In *Kitab Tahdid Nhayat al-Amakin*, he states:

> I say: if by adducing rational proofs and true logical syllogisms, we can conclude that the world was created, and that the parts of its finite period, since its creation and existence, have had a beginning, we cannot by such proofs deduce the magnitudes of those parts, which will enable us to determine the date of the creation of the world.[21]

Al-Biruni points out that while Jews, Christians, Sabians, Magians, and others agree about marking events in relation to the "era" during which mankind was created, they do not agree in estimating the duration of that era. Al-Biruni clarifies that while they consider the day of Genesis 1:1, 2 as "the first day of the week in which the world was created," he rejects this Judeo-Christian interpretation of the Genesis because he believes that days of creation "cannot be measured by a day and night." The reason for his rejection of this belief is that the cause of day and night is the sun "rising and setting," and that "the sun and moon were created on the fourth day of the week." He asks, "How is it possible to imagine that these days are like the days of our reckoning!" Then he counters these Judeo-Christian claims by quoting from the Qur'an: "A day in the sight of thy Lord is like a thousand years of your reckoning" (Sura 22: 47) and "In a day the measure whereof is as fifty thousand years" (Sura 70: 4).

Al-Biruni clearly believes that "we cannot estimate that period [of the creation] with our method of reckoning, and it is unverifiable since the beginning of creation." He used the Qur'an and logic to support his argument. Moreover, al-Biruni stated a thousand years ago that God initially scattered the celestial bodies and that the time between creation and the present could be billions of years. He states:

> For it is quite possible that these (celestial) bodies were scattered . . . when the Creator designed and created them. If you then ask the mathematician as to the length of time, after which they would meet each other in a certain point, or before which they had met each other in that identical point, no blame attaches to him, if he speaks of billions of years.[22]

Ameer Ali, a contemporary historian, described the close of the tenth century as the darkest period in the history of Islam in its rejection of rationalism and science. In his words, a "heartless, illiberal, persecuting formalism dominated the spirit of theologians; a pharisaical Epicureanism had taken possession of the rich and an ignorant fanaticism of the poor; the gloom of the night was fast thickening, and Islam was drifting into the condition into which ecclesiasticism had led Christianity."[23] During this epoch of travail and sorrow for all lovers of truth, a small body of thinkers joined each other to form a brotherhood aimed at keeping aflame the torch of knowledge among Muslims. They called themselves Ikhwan al-Safa or the Brethren of Purity. We know little about their membership with the exception of the following: Abu Sulayman al-Busti (known as al-Muqaddasi), Abu'l-Hasan al-Zanjani, Abu Ahamed al-Nahrajuri (known as al-'Aufi, and Zaid ibn-Rifaa.[24]

The tenth-century *Epistles of Ikhwan al-Safa* (*Rasa'il Ikhwan al-Safa*) is not the work of any one member but a collaborative effort. The *Rasa'il* is comprised of fifty-two epistles in which knowledge is classified into three classes: mathematics, physics, and metaphysics. The brethren maintained that the earth was created a very long time ago, and therefore has a long history. They tell us that the individual cycles of the stars affect events on earth and that it takes thirty-six thousand years for all the heavenly bodies to complete one cycle.

Some astronomical revolutions and conjunctions *Rasa'il Ikhwan al-safa* describes occur only once in a long time, while others repeat more frequently. "A very long period is that of revolution which occurs only in 36,000 years." Every three thousand years the fixed stars, apogees, and nodes of the planets pass from one astrological sign to another after having traversed all the degrees of these signs. Every nine thousand years they pass from one quadrant to another, and as the effect of their rays at diverse points of the earth are modified, so are the climates of different countries. Every thirty-six thousand years the stars conjunct and the planets reunite at the first degree of the sign of Ram to begin a new cycle.[25]

If it takes thirty-six thousand years to make one cycle, clearly Muslims accept that universe (and the Earth) is much older than the Biblical six thousand years.

Abu Ali ibn Sina (Latin name: Avicenna) (980-1037 A.D.) is "the most famous and influential of the philosopher-scientists of Islam."[26] His theory about the origin of mountains confirms the ancient age of the earth. He gives two causes for the emergence of mountains:

> Either they are the effects of upheavals of the crust of the earth, such as might occur during a violent earthquake, or they are the effect of water, which, cutting for itself a new route, has denuded the valleys, the strata being of different kinds, some soft some hard. The winds and waters disintegrate the one, but leave the other intact. Most of the eminences of the earth have had this latter origin. It would require a long period of time for all such changes to be accomplished, during which the mountains themselves might be somewhat diminished in size.[27]

Ibn Sina's statement, "It would take a long time for all such changes," is another proof that Muslims never entertained the Judeo-Christian belief of a young earth.

If we are to believe that the earth is only six thousand years old, as inferred in Biblical genealogies, then we should declare Darwin's theory invalid.[28] Charles Darwin could not reconcile the theory of evolution with the Biblical view of a six-thousand-year-old Earth, so he gradually drifted away from the Christian belief and become a materialist. He found compatibility between the theory of evolution and the idea of an antique Earth as stated in Charles Lyell's *Principles of Geology*. Hence, principles of uniformitarianism presented by Lyell should be included in any discussion of the theory of evolution.

The West claims that it was Charles Lyell who first declared that "most of the earth's rocks were formed over a long period of time by natural processes observable to this day"[29] and that he had arrived at this conclusion based on James Hutton's four principles of uniformitarianism:

(a) *The uniformity of law* states that the laws of nature that operate today operated in the past. The assumption about the unvarying natural law serves as a necessary warrant for extending inductive inference into an unobservable past.

(b) *The uniformity process* necessitates explaining the past through causes that are now in operation.

(c) *The uniformity rate, or gradualism* indicates that the pace of change is usually slow, steady, and gradual and that localized catastrophes such as floods, earthquakes, etc., occur.

(d) *The law of the uniformity, or non-progressionism* states that the earth operates in a dynamically steady state. We can infer its past not only from its laws and rate of change, but also from its current order of events. Land and sea might change their positions, but they exist through time in roughly the same proportion. Species come and go, but the mean complexity of life remains forever constant.[30]

The question is whether Hutton and Lyell were the earliest men who understood these principles of uniformitarianism.

In spite of the "ballads" written by Stephen Jay Gould and others in the West praising these men for the so-called discovery of the principles of uniformitarianism, a study of the works of the Muslim scientists—Ibn Sina, al-Biruni, Ikhwan al-Safa, Ibn-Khaldun, and others—reveals that these Muslim scientists understood and introduced these principles long before Hutton and Lyell.

Al-Biruni, the foremost Muslim geologist, was a uniformitarian. He identified the Ganges Plain in India as sedimentary deposit. He wrote:

> If you see the soil of India . . . if you consider the rounded stones found in the earth, however deep you dig, stones that are huge near the mountains and where the rivers have a violent current, stones that are smaller size at a greater distance from the mountains and where the streams flow more slowly, the stones that pulverized in the shape of sand where the streams begin to stagnate near their mouths and the sea . . . you can scarcely help thinking that India was once a sea, which by degrees has been filled by the alluvium of the streams.[31]

Moreover, al-Biruni wrote:

> We have to rely upon the records of rocks and vestiges of the past to infer that all these changes should have taken place in very, very long times and under unknown conditions of cold and heat: for even now it takes a long time for water and wind to do their work. And changes have been going on and observed within historical time.[32]

Al-Biruni did not invent extinct or unknown causes as explanations for the formation of the Ganges Plains in India. We infer from his statement "however deep you dig" that he used the scientific method to reach his conclusion. We see that he uses forces that operate today to explain changes in stones (the uniformity of law and uniformity process) and that he believed in the uniformity rate because he states, "by degrees has been filled by the alluvium of the streams." Furthermore, he recommended the so-called modern method of using rocks and fossils to decipher the long history of the earth.

In spite of this affirmation of the gradual and slow character of natural processes acting on the surface of the earth, al-Biruni knew that natural cataclysms and disasters devastated the earth from time to time. Like most other Muslim scientists, he believed that these catastrophes occurred in localized areas. According to al-Biruni:

> The disasters, which from time to time befall the earth, both from above and below, differ in quality and quantity. Frequently [the earth] has experienced one so incommensurable in quality or in quantity or in both together, that there was no remedy against it so that no flight or caution was of any avail. The catastrophe comes on like a deluge or an earthquake, bringing destruction either by breaking of the surface, or by burning by hot stones and ashes that are thrown out, by thunderstorms, by landslides, and typhoons; further by contagious and other diseases, by pestilence, and more of the like. Thereby a large region is stripped of its inhabitants, but then after a while, after the disaster and its consequences have passed away, the country begins to recover and to show new signs of life, then different people flock together like wild animals, who formerly were dwelling in hiding-holes and the tops of the mountains. Assisting each other against common foes, wild beasts of men, and furthering each other in the hope for a life in safety and joy civilizes them. Thus they increase to great numbers; but then ambition, circling round them with wings of wrath and envy, begins to disturb the serene bliss of their life.[33]

The Universal Flood of the Judeo-Christian faith is not a Muslim doctrine, and with the aid of scientific method, al-Biruni rejected a worldwide flood of the Judeo-Christian description. Muslims do believe it was a localized flood, however. Eight hundred years before Darwin, al Biruni stated:

> Persians, and great mass of the Magians, deny the Deluge altogether; . . . In denying the Deluge, the Indians, Chinese, and various nations of the east, concur with them. Some, however, of the Persians admit the fact of the Deluge, but they describe it in a different way from what is described in the books of Prophets. They say partial Deluge occurred in Syria . . . but did not extend over the whole of the then civilized world, and only few nations were drowned in it . . . the traces of the water of the Deluge, and effects of the waves are still present on these two pyramids halfway up, above which the water did not rise.[34]

In the above quotes, al-Biruni uses natural causes only to explain the destruction and recovery of people and civilizations. Unlike modern scientists, earlier Muslims believed that physical and chemical laws instituted at the time of the initial creation by God govern these natural events. Al-Biruni did not believe in the fixity of species as Lyell proposed in *Principles in Geology*, but in the changing complexity of life forms. His views on the evolution of life will be presented later with the pre-Darwinian Muslim thoughts on the origin of life and its development.

Ikhwan al-Safa (the Brethren of Purity) were also uniformitarians in their outlook. They believed:

> During the lapse of time of this movement [of stars], civilization in this world of generation and corruption is transported from one quarter of the earth to another. Continents replace seas, and seas come to replace the solid earth. Mountains change into seas, seas into mountains [i.e., the law of the uniformity, or nonprogressionism] . . . by the effect of their intermediate cause, the zeniths of the stars and the

incidence of their rays at diverse points of the earth are modified and with them the climate of diverse countries . . . . That is why the diverse regions of the earth are modified [i.e., the uniformity of law, the uniformity process, and the uniformity rate, or gradualism]. The layers of the air are changed above diverse places and countries by which the properties of these layers of air pass from one state to another. It is for this reason that cultivated earth becomes deserts, and deserts become cultivated earth; steppes become the sea and the sea becomes steppes or mountains.[35]

Again we find in the words of Ikwan al-Safa all the components of uniformitarianism. Unlike Lyell, they believed in a transformation of species and kingdoms to more complex life forms. We will learn about their point of view about life later.

Ibin Sina's (Avicenna's) view on the origin of mountains shows that this Muslim was also a uniformitarian.

Mountains have been formed by one (or other) of the causes of the formation of stones, most probably from the agglutinative clay, which slowly dried and petrified during ages of which we have no record. It seems likely that this habitable world was in former days uninhabitable, and indeed, submerged beneath the ocean. Then, becoming exposed little by little, it petrified in the course of ages, the limits of which history has not preserved; or it may have petrified beneath the waters by reason of the intense heat confined under the sea. The more probable of these two possibilities is that petrifaction occurred after the earth had been exposed, and that the condition of the clay, which would then be agglutinative, assisted the petrifaction.[36]

Ibn Sina observes that mountains most probably were formed from agglutinative clay, which slowly petrified "during ages of which we have no record" (i.e., the uniformity rate, or gradualism). We saw earlier that Ibn Sina attributed the flow of water and the earthquakes to the origin of the mountains (i.e., the uniformity process). He assumed these causes acted in the past as they acted during his time (i.e., the uniformity of law).

Ibn Khaldun, statesman, jurist, historian, and scholar, was born in Tunis on May 27, 1332. The love of learning and intellectual pursuit for which his father and grandfather were noted produced in him the rare combination of philosopher and statesman. Ibn Khaldun served as advisor to various rulers of his time. In 1387 he made the pilgrimage to Mecca, after which he returned to Egypt, where he died on March 17, 1406.

His well-known work, *Muqaddimah* (*Introduction*), is the preface to *Kitab al 'Ibar*, which is considered the first attempt to interpret the pattern of changes that govern developments in the social and political order of societies. A rational analytical methodology and an eye for detail marked his approach. Ibn Khaldun diverged from historical narration of events based on prevailing social conventions. He created a philosophy of history that was detached from superstition and based on a critical and scientific examination of events. Thus, Ibn Khaldun pioneered a new science and new terminology to the study of history.[37] Arnold J. Toynbee, an eminent British historian, believed that the *Muqaddimah* was the "greatest work of its kind that has ever been created by any mind in any time."[38]

Ibn Khaldun informs us that Christian genealogists used to believe that black Africans were the children of Noah's son Ham and that their blackness resulted from the curse Noah cast on them. He points out that the story in the Torah declares that Ham's children would be slaves to his brother's descendants, but makes no mention of their color. In his view, attributing the blackness of Africans to Noah's curse discounts the influence that natural forces such as warm and cold climates have on living creatures. He explains:

> Genealogists who had no knowledge of the true nature
> of things imagined that Negroes were the children of
> Ham, the son of Noah, and they were singled out to be

black as a result of Noah's curse, which produces Ham's
color and slavery of God inflicted upon his
descendants . . . . The black skin common to the
inhabitants of the first and second zones [Ibn Khaldun
divided northern hemisphere of the earth into seven
north zones. The first zone runs along equator and is
followed to the north successively by the second through
seventh zones.] is the result of the composition of air in
which they live, and which comes about under the
influence of the greatly increased heat in the south. The
sun is at the zenith there twice a year at short intervals.
In all seasons, the sun is in culmination for a long time.
The light of the sun, therefore, is plentiful. People there
go through a very severe summer, and their skins turn
black because of excessive heat.[39]

Ibn Khaldun attributes the whiteness of the inhabitants of
northern zones to the influence of excessive cold climates that
occur as a result of the greater distance between the sun and the
zenith in those regions. He also attributes to climatic conditions
northern people's lack of bodily hair, their blue eyes, freckled skin,
and blond hair, and warns that Christians should acknowledge
the influence of geographic and environmental factors that create
different features among people. He considered it a mistake for
Christians to attribute the traits of whole nations to one of their
prominent ancestors. Ibn Khaldun also contends that:

[There also is disregard of the fact that physical
circumstances and environment] are subjected to
changes that affect later generations; they do not
necessarily remain unchanged. This is how God proceeds
with His servants . . . . And verily, you will not be able to
change God's ways.[40]

Ibn Khaldun applied the principles of uniformitarianism when
he attributed the composition of the human race to causes that

were in operation during his time (i.e., uniformity process). He observed that the hot and cold climate and the position of the sun in various zones of the earth are constant in time and place (i.e., uniformity of law and the uniformity rate, or gradualism). Lyell believed that species come and go, but the mean complexity of life remains forever constant. His opinion is similar to the then-held Judeo-Christian view. Edey and Johanson describe it as follows: "Because of a general acceptance of the Biblical account of Creation, almost everybody took it for granted that species were fixed. They have been created by God in their own shape and could not change."[41]

Ibn Khaldun and other Muslims who preceded Lyell did not believe in the story of Noah's curse that human traits are predetermined. On the contrary, they believed that human species do undergo change. Ibn Khaldun maintained that environmental changes "affect later generations." His earlier statement in which he refers to God as the instigator of change affirms the fact that Muslims in their classical period believed that the impact of environmental changes on life forms was the way by which God created new and varied life forms.

All four components of uniformitarianism are obvious in early Muslim thought, but they did not adhere to Lyell's idea of constancy in the complexity of life. Thus, neither the idea that environmental factors cause changes in life forms nor uniformitarianism are pristine Western ideas, but Muslim discoveries. These principles were widely accepted among early Muslims who routinely used them in scientific pursuit long before Lyell and Hutton.

Western historians of science, who have claimed virtually all scientific discoveries, have also taken credit for the rational explanation of the origin of fossils. Fossils are important strands of evidence supporting evolution. Most people in the Christian West did not know what to make of it. In the Dark Ages, the Church explained that these fossils were Satan's failed attempt to mimic the work of God in creating life.

Asimov attributes the first reliable rationalization of the origin of fossils to Leonardo da Vinci (1452-1519), who claimed that they were remnants of obsolete creatures. According to Asimov, Charles Bonnet (1720-1792), in an attempt to avoid an explanation that is based on the concept of evolution, promoted the belief that fossils were relics from the time of Noah's flood. In 1770 he spread the concept of catastrophism, by which he held that the entire living creatures on the Earth was destroyed by whole series of catastrophes and then a new creation began. Bonnet argued that the Bible dealt only with the Earth after the last catastrophe. As Asimov observes, however, the discovery of more fossils required the recruitment of more and more catastrophes, which led to the failure of the concept of catastrophism. The origin of fossils increasingly began to be seen more as a process of evolution than the product of catastrophes.[42]

Asimov's statement that no one knew the origins of fossils and that da Vinci was the first to render a rational explanation about their origin is another Western myth. Long before da Vinci, the early Muslims knew that fossils were the remains of once-living organisms. Al-Biruni gave his opinion about the origin of fossils in his description of the Arabian Peninsula:

> This steppe of Arabia was at one time sea, then was upturned so that the traces are still visible when wells and ponds are dug; for they begin with layers of dust, sand and pebbles, then there are found in the soil shells, and bones, which cannot possibly be said to have been buried there on purpose. Nay, even the stones are brought up in which are embedded shells, cowries and what is called 'fish-ears', sometimes well preserved, or the hollows are there of their shape while animals have decayed.[43]

Similarly, Ibn Sina's opinion about fossils can be derived from his description of the origin of mountains. He wrote:

> It would require a long period of time for all such changes to be accomplished, during which the mountains themselves might be somewhat diminished in size. But

that water has been the main cause of these effects, is
proved by existence of fossil remains of aquatic and
other animals on many mountains.[44]

Important scientific discoveries rendered by Muslim savants
about the age of the earth, uniformitarianism, and the origin of
fossils, have been almost completely obliterated from the sight of
the world by Western scholarship. Western historians generally
start with the Greeks and skip over Islamic history to the
Renaissance in Europe. Their amnesia about the contribution of
Islamic scientists clearly presents a distorted version of history.

Why aren't the contributions of Muslims widely known?
Professor John William Draper (1812-1883) of New York
University and a contemporary of Darwin, points to persistent
efforts by Western scholars to suppress such history:

> I have to deplore the systematic manner in which the
> literature of Europe has contrived to put out of sight
> our scientific obligations to the Mohammedans. Surely
> they cannot be much longer hidden. Injustice founded
> on religious rancor and national conceit cannot be
> perpetuated forever.[45]

Draper's last statement clearly attributes the erasure of Muslim
contributions from the Western historical record to religious and
racial prejudices. In *The Making of Humanity*, British historian
Robert Briffault focuses on the deficiencies of the Western historical
record by demonstrating that the European Renaissance occurred
"under the influence of Arabian and Moorish revival culture [and
that] Spain, not Italy, was the cradle of the rebirth of Europe." He
explains that while Europe wallowed in ignorance and barbarism,
Muslim cities constituted "centers of civilization and intellectual
activity." He asserts: "[T]hat a brilliant and energetic civilization full of
creative energy should have existed side by side and in constant relation
with populations sunk in barbarism, without exercising a profound
and vital influence upon their development, would be a manifest

ɴistorians to restrict mention of such influence
Ϲross over Crescent" and "the reclamation of
yoke" does not, in his view, erase the relationship
ɪc culture and Europe, despite "the conspiring of
ɀe to suppress, deform, and obliterate the records of
thaᴜ

Briffaᴜ. ɔhoes Draper's accusation of the scholarly suppression of the truth that was "stubbornly ignored and persistently minimized" even by prominent historians such as Gibbon. Not only does Briffault consider the past representations of Islamic culture of the Middle Ages as inaccurate, but also those of the present. Briffault's resentment of the scholarly injustice toward the Islamic culture led him to the following insight: "[I]t is highly probable that but for the Arabs modern European civilization would never have arisen at all."[47]

Briffault's statements are evidence of scholarly truthfulness, an effort that almost always requires a degree of ethical courage, and without which scholarly studies will, as times change, lose their credibility. How fitting it is to quote, in this instance, an old African proverb about history and historians: "Until the lions have their historian, tales of hunting will always glorify the hunter."

The story of science has often been presented as the victory of scientific discovery over religious mythology. But another story is waiting to be told. The question is whether the East or the West, the Christians, Jews, Muslims, and others, have enough intellectual freedom to listen to that forgotten story.

CHAPTER 5

A Brief Review of the Theory of Evolution

I have chosen to offer a brief review of the theory of evolution for the benefit of those who are unfamiliar with the field. This overview is also essential to an understanding of where the Qur'an agrees with the theory of evolution and where it does not.

The Christian fundamentalists present the theory of evolution as anti-God, primarily to the American public but also to Christians worldwide. Recently some Muslims have adopted the hollow Christian fundamentalist polemics to discredit evolution from their pulpits as well. In this chapter we will discuss antievolutionists' ardent outcries against evolution, and evolutionists' patient response.

*   *   *

A bush takes life from a seed and grows into a stem, branches, twigs, leaves, and fruit. Similarly, life originated from organic and inorganic chemical matter, grew into an original "stem cell," and then branched into every kind of life form on the earth. Biologists have constructed a model to explain the origin of life and its growth, a process that can be likened to the development of a bush. This model is the theory of evolution.

When those who are not biologists talk about the theory of evolution, however, they are often confused about the words theory and evolution. The Institute for Creation Research (ICR) at San Diego, the foremost American fundamentalist Christian creationist organization, reinforces the innocent confusion that is prevalent among the general public. Many Muslims also use this confusion to reject evolution. Therefore, we need to explain the scientific meaning of these words.

Professor Stephen Jay Gould of Harvard University, a leading advocate for the theory of evolution, explains the word theory as follows: "In American vernacular, 'theory' often means 'imperfect'—part of a hierarchy of confidence running downhill from fact to theory to hypothesis to guess." Anti-evolutionists therefore argue that evolution is a theory and not a fact, and that it is inferior to fact.

Gould clarifies: "[Evolution] is a theory. It is also a fact, and facts and theories are different things, not rungs in a hierarchy of increasing certainty. Facts are the world's data. Theories are structures of ideas that explain and interpret facts." Theories connect all facts together to form an inherently consistent story. Gould points out that: "Facts do not go away when scientists debate rival theories to explain them." According to science "'fact' can only mean 'confirmed to such a degree that it would be perverse to withhold provisional assent.'"[1]

In short, the fact of evolution does not go away when varying theories are offered to explain the process.

Charles Darwin did not use the word *evolution,* but *evolve* appeared as the last word in his work, The Origin of Species. *Webster's Third International Dictionary* describes the word *evolution* as "a process of continual change from a lower, simpler, or worse condition to a higher, more complex, or better state." *The American Heritage Dictionary* explains evolution as "a gradual process in which something changes into a different or better form." Darwin himself understood the word as described in the dictionary.

The contemporary scientific definition of the word *evolution* differs greatly from Darwin's understanding and what is given in standard dictionaries. When common people talk about the theory of evolution, however, they tend to use the meaning given by a standard dictionary.

Scientists say that these definitions are simply wrong. Standard biology textbooks define evolution thusly:

> Evolution can be precisely defined as any change in the frequency of alleles within a gene pool from one generation to the next.[2]

A gene pool is the set of all genes in a species or population. For geneticists, a population means a group of organisms that interbreed or have the potential to do so. Alleles are alternative variants (different versions) of the same gene. For example, humans can have A, B, or O alleles that determine one aspect of their blood type. For scientists, the genetic composition of a certain population exists and evolves over time.

Let us use a frequently cited example of the *Biston betularia* (the peppered moth) to explain evolutionary change. These moths are either light or dark. Primarily, a single gene determines the color of moths. The dark moths constituted two percent of the population prior to 1848, then gradually increased until the 1950s. Hence in the moth population, there was an increase in the presence of the gene that gives the moth dark color between 1849 and the 1950s. The change in the frequency of dark moths represents a change in the gene pool. Scientists call this change evolution.

According to Robert Shapiro, professor of chemistry at the University of New York, the origin of life remains still a great mystery and we do not know how life began on earth.[3] However, the general scientific view is that life somehow evolved through a series of chemical reactions.

In the 1920s Alexander Oparin in the Soviet Union and J.B.S. Haldane in England independently introduced ideas that explained the origin of life on Earth; later other scientists developed a comprehensive theory that has become the cornerstone of scientific explanation of the origin of life. They maintain that in the past, the atmosphere of the earth was much different than it is today. After the earth and the solar system were formed out of the condensation of interstellar materials, the atmosphere had an abundance of hydrogen, methane (a simpler reduced form of carbon), and ammonia (a simpler reduced form of nitrogen). New compounds were formed as a result of the exposure of the early atmosphere to lightning, radiation, volcanoes, and meteors. Rainfalls would then siphon these compounds into oceans and ponds, which are sometimes referred to as "pre-biotic soups." The chemicals in the pre-biotic soup combined and transformed into more complex compounds and ultimately into life itself.

Stanley Miller and Harold Urey in the 1950s provided laboratory evidence for the formation of a number of organic compounds, including several amino acids, by exposing the mixture of hydrogen, ammonia, and methane to electric sparks for several days. Amino acids are important components of deoxyribonucleic acid (DNA)[4] (Figure 5-1).

Miller and Robert Orgel assumed that "there must have been a period when the earth's atmosphere was reducing (without free oxygen), because the synthesis of compounds of biological interest takes place only under reducing conditions."[5] The Miller-Urey experiment has been used to support the Oparin-Haldane theory of the origin of life.

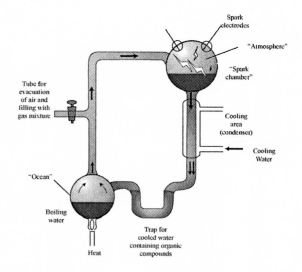

Figure 5-1. MILLER-UREY EXPERIMENT SIMULATING PRE-BIOTIC CONDITIONS. The upper glass chamber ("early atmosphere") has a mixture of hydrogen (H_2), methane (CH_4), and ammonia (NH_3). Sparks from electrodes represent lightning as the source of energy. The lower glass chamber ("ocean") is boiled to make water vapor, which provides oxygen to the chemical reaction and also carries organic molecules back to the lower chamber ("ocean") by condensation ("rainfall").

If, as Miller and Urey claim, the production of organic compounds depends on the presence of a reducing atmosphere (lack of free oxygen in the environment), then the earth's initial atmosphere must have been reducing to form organic compounds. Recent data points to the absence of such a reducing atmosphere in the past. Professor Shapiro of New York University tells us that, according to geologists and geochemists, the earth's atmosphere was initially neutral or "slightly reducing," and methane and ammonia would not be formed. In addition, radiation from the sun, which recent discoveries have found to be more intense in the past, would have destroyed any reduced gases.

Also contrary to the Miller-Urey claims, no extremely nitrogen-rich carbonaceous deposits out of which pre-biotic soup could have been formed have been discovered. Moreover, in 3.8-billion-year-old rocks in Greenland we see carbon in oxidized carbonate form. Shapiro points to the paradox that the earth, the place where life originated, is a "non-reducing oasis in [a] reducing universe." He concludes that either science has misinterpreted the conditions under which life was generated or the reducing atmosphere happened in a different way.[6]

Philip H. Abelson of the Geophysical Laboratory at Carnegie Institute of Washington maintains that there is no evidence to support the methane-ammonia hypothesis and that if ammonia existed, it would have "quickly disappeared."[7] Erich Dimroth of the University of Quebec and Michael M. Kimberley of Toronto University argue against a reducing atmosphere in the earth's past. They confirm the fact that we find "no evidence in the sedimentary distribution of carbon, sulfur, uranium, or iron that an oxygen-free atmosphere has existed at any time during the span of geological history recorded in well preserved sedimentary rocks."[8]

Henderson Sellers of the University of Liverpool, and A. Benlow and A. J. Meadows of the University of Leicester express a similar opinion. They explain their view as follows:

> The mean temperature of the early Earth was above 273 K, even though the solar luminosity was low and

significant amounts of NH_3 were not present in the atmosphere. Biologists concerned with the origin of life still often quote an early atmosphere consisting of reduced gases, but this seems to stem as much from ignorance of recent advances as from active opposition to them. In the later part of the 1970s, the concept of early-oxidized atmospheres on the terrestrial planets is becoming the new orthodoxy.[9]

After reviewing the concept of the atmospheric state of the primitive earth, Shapiro finds unscientific the acceptance of many scientists of a theory as fact and their shunning of any evidence that contradicts it.[10] Even though there remains much dispute about the origin of life in the earth's early atmosphere, there is no disagreement that amino acids are the building blocks of Deoxyribonucleic acid (DNA) and that DNA is the basis of all life on the earth.

After the formation of amino acids, reactions caused by further condensation led to the development of nucleotides. A nucleotide is composed of one five-carbon sugar (deoxyribose), one phosphoric acid radical, and one nitrogen base out of adenine, cytosine, guanine, or thymin. Repetition of four units of nucleotides forms a unit of DNA, the basic building block of life (Figure 5-2). Most scientists admit that the "spontaneous" formation of small molecules such as amino acids and nitrogen bases could have occurred under various earthly conditions in the past; however, they are not certain how these molecules could have merged into larger units.[11]

Even though scientists believe that life began in water, until recently there has been no sound explanation for the paradox of polymerization in the presence of water. Polymerization is a chemical reaction that joins small molecules into large chains of molecules. The product formed after polymerization is known as a polymer. Polymerization reactions are generally dehydrations in which a molecule of water is lost in the formation of a polymer. The polymer is unstable in the presence of water.

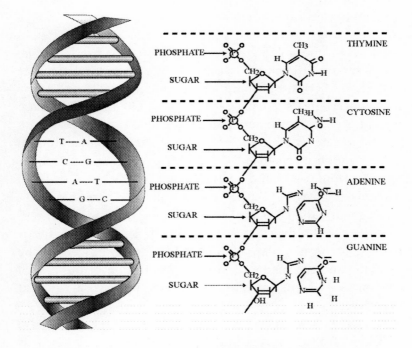

Figure 5-2. PORTIONS OF SINGLE DNA POLYNUCLEOTIDE CHAIN. Each nucleotide is composed of one five-carbon sugar (deoxyribose), one phosphoric acid residue, and one nitrogenous base (adenine, cytosine, guanine, or thymine). The chain also consists of alternating phosphate and sugar residue.

Shapiro cites A.G. Cairns-Smith, a chemist at the University of Glasgow in Scotland, who maintained that primitive forms of life on earth were made of clay minerals. Cairns-Smith explains that the complex composition of clay minerals facilitates the catalysis of chemical reactions.[12] According to a famous British scientist, J. D. Bernal, the attachment of molecular intermediates on clay minerals may have drawn these intermediates together into high concentration. Such a concentrating process could check the tendency of water to break down the polymers necessary for the evolution of life. The catalysis of chemical reactions by clay mineral suggests that organic synthesis could also occur in deep water where ultraviolet light is nonexistent. This would help to explain how phosphates could have been integrated preferentially into pre-biological organic molecules at a time when biological concentration mechanisms did not yet exist.[13]

Therefore, as some scientists claim, clay played a role to condense the sugar, phosphates, and nitrogen base into DNA. Subsequently the DNA became more complex, and organized, through various stages, into a living cell. The clay hypothesis has yet to be demonstrated in a laboratory, however. The Qur'anic and the medieval Muslim description of the role of clay minerals in the creation process are discussed in the next two chapters.

Another scientific claim, by Manfred Eigen, William Gardiner, Peter Schuster, and Ruthild Winkler-Oswatisch, is that the earliest genetic unit was ribonucleic acid (RNA). They assumed that RNA preceded DNA because the function of the complex DNA structure depends upon RNA. Thus, they presumed that Life on the earth began with the accidental formation of replicating RNA molecules and that "after a period of competition and evolution, a winner emerged."[14] Eigen's theory posits that RNA molecules (by themselves) rearranged their structure and somehow directed themselves to a synthesis of more complex enzymes, which led to "accidental" development of life.

Maitland Edey, editor of Time-Life Books, and Donald C. Johanson, director of the Institute of Human Origins, critique Eigen's theory. They find it "odd to speak of things like 'selection,' 'competition,' and 'success' taking place among the inert chemicals." In their view Eigen's model is weak because "aside from the initial step, none of the next ones have been shown to have happened." Moreover, they point out that Eigen's mathematical calculations are flawed.[15]

It may seem odd for Edey and Johanson as well as other materialists to speak of competition between "inert chemicals." However, within the Islamic theory of creation described in this work, we will see how such competitions between all so-called "inert chemicals" and all big or small components of the universe mesh with Allah's process of creation.

According to evolutionists, the primitive cell gradually evolved into plants and animals. This evolution was due to spontaneous mutation or changes in the genetic material within the cell. Scientists believe that all life forms are changing and that humans may evolve into another form, the nature of which is not known to us. Spontaneous mutation and the idea of the possibility of humankind evolving into another species is nothing new within the medieval Islamic concept of creation. These points are discussed later in this book.

In *The Origin of Species,* Charles Darwin proposed that in nature, there is a struggle for survival between organisms, which he calls natural selection and some scientists call survival of the fittest. He described the outcome of this struggle as follows:

> Owing to this struggle, variations, however slight and from whatever cause proceeding, if they be in any degree profitable to the individuals of a species, in their indefinitely complex relations to other organic beings and to their physical conditions of life, will tend to the preservation of life and will tend to the preservation of such individuals, and will generally be inherited by the offspring. The offspring, also will have better chance of surviving, for, of the many individuals of any species

> which are periodically born, but a small number can
> survive . . . from the war of nature, from famine and
> death, the most exalted object which we are capable of
> conceiving, namely, the production of the higher animal
> directly follows.[16]

Darwin concludes that "there is grandeur in this view of life, with its general powers having been originally breathed by the Creator into few forms or into one; and that, whilst this planet has gone cycling on according to the fixed laws of gravity, from so simple a beginning endless forms, most beautiful and most wonderful have been, and are being evolved."[17]

Later in this book we will see that the idea of evolution was well accepted by Muslims centuries before Darwin, and they describe it almost verbatim as Darwin describes here.

G. H. Hardy and W. Weinberg, the fathers of evolutionary mathematics, explained the mechanism of evolution. They state that Mendel's laws of inheritance predict the outcome of single-pair mating but do not predict the proportions of genotypes in a population. A genotype is the genetic constitution of (or the totality of genes possessed by) an organism (or a group) irrespective of its physical appearance.

The genotype reflects the frequency of alleles in the population. Alleles are the pairs of alternative forms of gene, one from each parent, for a particular Mendelian character—for example, black or white complexion—occupying same position in a homologous chromosome. Hardy and Weinberg find that both allelic and genotypic frequencies remain unchanged from generation to generation unless their equilibrium is disturbed by external factors such as natural selection, mutation, migration, or genetic drift.[18]

An example of natural selection that disturbs the equilibrium is the change in the frequency of the gene for melanin in our friend *Biston betularia*. Before industrialization in the United Kingdom, the dark-colored peppered moths with melanin genes were fewer in number than light-colored moths. Environmental pollution blackened the barks of trees where the moths rest in the daytime, allowing the dark-colored insects to blend into the darkened barks. The light-colored moths became more obvious prey for the birds. This led increased survival and reproduction of the moths that carried the melanin gene. In the 1950s, when a significant decrease in the industrial pollution occurred, the light-colored moths again began to grow in number. This transformation at the subspecies level is a well-documented example of natural selection caused by external factors within historical time.

Other than natural selection, the migration of organisms to other host populations affects allelic and genotypical frequencies as a result of mating between the migrant and the host populations. This is known as gene flow, as it causes changes in the gene pools. Such changes take place, for example, in plants when the wind spreads seeds or pollen beyond the bounds of a local plant population.

Still another important factor that alters the allelic frequency is genetic drift. When a change in the proportions of alleles in a gene pool occurs by chance, various outcomes can result. The alleles, which form the gene pools of the next generation, are samples from those of the current generation.

Sampling new alleles can be compared to sampling marbles from a bag filled with nine hundred black marbles and one hundred white marbles (90% black and 10% white). If someone picks up ten marbles at random without looking into the bag, there is a chance that he will get no white marbles. Thus, pure chance can cause deviations in the expected frequency of white marbles. Such random change in the gene pools of a particular population is called genetic drift.

As in the marble sampling case, random selection operates in the propagation of alleles from parent population to offspring population. Human beings have no way of predicting the exact constitution of an individual offspring, but they can predict the gene pool of a certain population if mutation is excluded. When we look at the fertilization process, organisms form many more gametes than those actually taking part in such fertilization. There is always a chance that an allele will be missing from those that contributed to the genetic constitution of the next generation, primarily due to sampling error (random selection of one gamete out of a few million that fertilize one ovum). Thus, the gene pool changes by chance alone.[19]

For Christian fundamentalist creationists and some Muslims, random events are chaotic and purposeless. As they use the words *random* and *chance* to discredit science and the theory of evolution, we should briefly clarify the scientific meaning of these words. The technical meaning of random is that occurrences are governed by equal probability. When Scientists commonly speak of the chance of something happening, it means that it will occur according to a known probability. For example, the chance of getting a head in a coin toss is 1 in 2. Similarly, by knowing the number of throws, we can predict the number of times the dice total seven. When the number of tosses becomes quite large, it is possible to estimate and predict ranges of error for the frequencies of heads or the number of sevens in a row accurately. This is all based upon mathematical formulas from the probability theory. Therefore, a natural phenomenon that is governed by chance yields "maximal simplicity, order, and predictability." Thus, the randomness in nature and the probability theory becomes the best tool to study and understand long historical sequences, such as the history of life.[20]

To discredit the theory of evolution, "scientific" creationists argue that the human eye defies evolution. They posit that the complex system of lens, iris, photoreceptors, and special nerves of the eye cannot be explained by the randomness of evolution. They say that the odds of an eye evolving are equivalent to a monkey typing Shakespeare's *Hamlet* and could not happen by random chance.

Evolutionists counter the argument of so-called scientific creationists by asserting that natural selection is not random, but a trial and error process that "preserves gains and eradicates mistakes." They support their claim with examples:

> The eye evolved from a single, light-sensitive cell into the complex eye of today through hundreds if not thousands of intermediate steps, many of which still exist in nature. In order for the monkey to type the first 13 letters of Hamlet's soliloquy by chance, it would take 26 to the power of 13 number of trials for success . . . if each correct letter is preserved and each incorrect letter eradicated, the process operates much faster. How much faster? Richard Hardison constructed a computer program in which letters were selected for or against, and it took an average of only 335.2 trials to produce the sequence of letters TOBEORNOTTOBE. This takes the computer less than 90 seconds. The entire play (of *Hamlet*) can be done in about 4.5 days.[21]

Moreover, we have evidence that the eye may have evolved independently many times during the history of life. It progressed from a simple eye spot made of light-sensitive cells, as in the flat worm, to individual photosensitive units, as those in insects with light-focusing lenses, to the eventual form of an eye with a single lens focusing images onto a retina. In man and other vertebrates, the retina consists not only of photoreceptor cells, but also of neurons that analyze the visual images. Through gradual evolutionary steps, different kinds of eyes have evolved, from simple light-sensitive organs to highly complex systems of vision.

The fact that evolution has taken place is no longer a disputed issue among the majority of biologists. However, as D. J. Futuyma, professor at State University at New York, tells us, theoretical explanations of the causes for evolutionary change remain controversial.[22] Two predominant theories explain the patterns of evolution: the neo-Darwinian evolutionary synthesis, which Ernest Mayr of Harvard University calls phyletic gradualism, and Gould's punctuated equilibria.

The modern evolutionary synthetic theory holds that small mutations take place in an organism, resulting in small variations—

in size, strength, shape, behavior, intelligence, endurance, and so on—among the new generations. Ecological factors include natural selection screening out some combinations of mutant genes to form subspecies. Isolating mechanisms, such as geographic barriers, habitat preference, seasonal breeding, behavioral (ethological) differences, etc., prevent the subspecies (incipient species) from merging with the general populations from which they originated. Thus, natural selection aided by the isolating mechanisms produces good species.[23] In the words of Ernest Mayr, the evolutionary synthetic theory, or "gradual evolution . . . [is] explained in terms of small genetic changes (mutation) and recombination, and the ordering of this genetic variation by natural selection."[24]

The implication of classical Darwinism and neo-Darwinism is that the fossil record should reflect slowly changing forms— that the intermediate forms of life should be preserved as fossils and we should be able to unearth them. Darwin wrote in *The Origin of Species:* "I look at the geological record as a history of the world imperfectly kept, and written in a changing dialect; of this history, we possess the last volume alone, relating only to two or three countries. Of this volume, only here and there a short chapter has been preserved; and of each page, only here and there few lines."[25]

Many more fossils have been discovered over the hundred years since Darwin's death, of course. We are no longer limited to the most recent volume, and some of the volumes are now far more complete. Nevertheless, fewer examples of gradual change have been found than might have been expected. Therefore, many evolutionary biologists have questioned the modern evolutionary synthesis theory. Steven Stanley, a paleontologist at Johns Hopkins University, points out that there is no record of phyletic evolution to validate gradualism.[26] Similarly, David M. Raup, geologist and curator of the Field Museum of Natural History in Chicago, maintains that "[I]nstead of gradual unfolding of life, what geologists of the present day actually find is a highly uneven or jerky record; that is, species appear in the sequence very suddenly, show little or no change during their existence in the record, then abruptly go out of the record."[27]

From Darwin's time until recently, discrepancies in the model of phyletic change in the fossil record have been ascribed to imperfections in the fossil record itself. Other causes are postulated to be erosions, lack of proper conditions for fossilization, or simply not looking at the right place.

Nils Eldredge and Stephen Jay Gould have a different explanation for the morphological invariance of species through time and the unconnected intermediates in the fossil record. The theory of punctuated equilibrium by Eldredge and Gould explains the rarity, not absence, of transitional forms. Gould posits punctuated equilibria as a sound explanation for the inconsistencies of the fossil record. He dismisses the idea of gradual evolution and affirms the idea of evolution that is more comparable to "climbing a flight of stairs (punctuation and stasis) than rolling up an inclined plane."[28] New species usually appear by splitting off of small populations from an ancestral stock. There is no slow and steady transformation of entire ancestral population in this process. Stephen J. Gould and other biologists argue that the frequency and speed of the origin of most species by splitting happen over a range of hundreds or thousands of years.[29]

Some Christian and Muslim anti-evolutionists distort the theory of punctuated equilibrium by propagating the idea that this theory does not accept the existence of any transitional species. But punctualism clearly postulates the existence of morphological intermediates as small populations. Punctualists claim that evolution is continuous but jerky, with established lineages.[30] A species remains static without any morphological change. Then, a small peripheral isolated population of the species changes rapidly into a new species as a result of mutation in the regulatory gene that controls the expression of the structural gene.[31] The new species migrates into the parent species.

Gould and Eldredge have not claimed their explanation of gaps in the fossil record to be a new discovery. It is only a new interpretation for the paleontologicial observations of "the geologically instantaneous origination" and subsequent stability (often for millions of years) of paleontological species. [32]

For the anti-evolutionists among the Christians and Muslims, the phrase "geologically instantaneous" has become music for their ears. They proudly proclaim that Gould's descriptive phrase is the confirmation of their belief that God created all species abruptly from the thin air like a magician! It appears that they have a selective deafness. They hear only what they want to hear. Gould has pointed out many times that the instantaneous speciation is an artifact of the compression of time that takes place each time a new layer is formed in the fossil record. The geological moments are equal to millions of years of human time frame.

The study of life in the geological past is called paleontology. Sedimentary rocks are composed of sediments that gradually settled in cumulative layers at the bottom of oceans and large lakes, carrying with them the remains of plants and animals. The deepest strata of these rocks are the oldest, and the shallowest are the most recent.

Paleontologists have recovered and scrutinized the fossil remains of thousands of organisms that lived in the past from these sedimentary rocks, and they have arranged them according to the geological time scale. Fossil records show that many kinds of extinct organisms were different in form from those living now. They reveal successions of organisms, with simpler life forms appearing early and complex forms appearing later (Table 5-1).The stratigraphic dating, as described in Table 5-1, is reliable but not absolute. It shows only the chronological order for the appearance of organisms, but does not tell the exact time of their appearance within the chronological age of the earth.

Era	Millions of years before present	Biological Events
CEMOZOIC (AGE OF MAMMALS)	0.3	Modern man
	1.0	Homo erectus
	2.0	Homo habilis
	5.0	Australopithecus afarensis
	25.0	Ramapithecus
	37.5	Old world monkey
	55.0	New world monkey
	75.0	First placental mammals
MESOZOIC (AGE OF REPTILES)	135.0	Climax of reptiles,
	185.0	Reptiles dominant; First bird; first mammals. First flowering plants.
	225.0	First dinosaurs; first appearance of frogs, turtles, crocodiles, mammals, flying reptiles; conifers dominant.
PALEOZOIC	275.0	Widespread marine life extinction; mammal-like reptiles flourish.
	310.0	First reptiles appear; Giant insects on land; amphibians thrive in sea
	350.0	Echinoderms abundant in seas; terrestrial amphibians rare.
	395.0	First bony fishes; sharks.
	435.0	Earliest known land plant; Fishes become important.
	480.0	Earliest known fishes.
	600.0	Organisms with hard parts appear; fragmentary vertebrates appear.
PRECAMBRIAN	3000-4800	Soft bodied primitive life; algae; protozoans; molluscs; annelids; arthropods; others of uncertain affinities; bacteria.

Table 5-1. ORGANISMS IN GEOLOGICAL TIME SCALE.

Absolute dating is determined by radiometry, which uses the rate of decay of radioactive elements in the geological deposits and fossils to calculate their age.

Radioactive dating has shown that the earth was formed 4.5 billion years ago. The study of fossils clearly indicates that not all forms of life originated simultaneously, nor in six days. The earliest fossils of microorganisms resembling bacteria and blue algae are three to five billion years old. The oldest animal fossils are of small wormlike creatures. They are seven hundred million years old.

Animals with backbones, known as vertebrates, appeared as primitive fishlike organisms four hundred million years ago and gradually evolved into modern fish. Later, amphibians and birds appear in the fossil record. Finally, mammals appear in fossils of about two hundred million years old. (Mammals are a class of warm-blooded vertebrates who secrete milk to feed their young.)

The most primitive, single-celled organisms are found in the oldest and deepest rocks, while chronologically younger rocks contain progressively more complex forms of life.[33] Such a distribution and layering suggests evolution for many agnostics and creation by modification for many God-conscious people of all faiths.

Comparative anatomy is the study of inherited similarities in bone structure and soft tissues among various organisms. Even without fossil evidence, we are able to infer that organisms have evolved by adapting structures of other species that still survive today. The skeletons of bats, whales, humans, and horses are remarkably similar, even though they live in different environments and have very different lifestyles. Homologous versions of different bones are found not only in the limbs of the various animals, but throughout the remainder of their bodies as well. The various homologous bones enable humans to write, bats to fly, whales to swim, and squirrels to carry and hide their food for the coming winter. The concept that each individual species emerged separately from others is not congruent with the evidence of similar bone

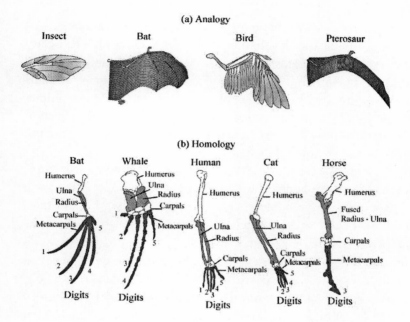

Figure 5-3. ANALOGY AND HOMOLOGY. (a) Analogy: Wings of a fly, a bird, and a bat. The wings of a fly, a bird, and a bat are analogous structures that are developed from different materials. Evolution of each was independent. (b) Homology: limbs of a bat, whale, human, and horse. Bones with the same intensity of shading are homologous throughout, even though they are modified in size and in shape by reduction or by fusion of bones. Some materials have been changed to meet the needs of different animals.

structures found across different types of animals. Similarities of skeletal anatomies can be explained as structures inherited from common ancestors and modified to adapt to the different environments and ways of life that each species had to deal with.[34] (Figure 5-3.)

Further evidence of common ancestral species can be gained from studying the development of animal embryos. The early embryos of vertebrates such as fish, birds, lizards, and mammals (including humans) are strikingly similar. That most animals develop gill slits below their heads suggests that these animal embryos may have descended from ancestral species that lived in the water, although many embryologists believe that these are not gills, but mere folds (flexures).

Scientists have found that across different species, structures analogous to the bones of the forehand all develop from the same region of the embryo. Moreover, vertebrate embryos initially have tails. In certain species (including humans), tails disappears as embryos mature, except in some extremely rare cases in which human babies are born with tails. This common developmental pattern across various species of animals and humans reflects our evolutionary kinship within the animal kingdom.[35] (Figure 5-4.)

Vestigial organs are structures that are greatly reduced, often without any function in the body. Such vestiges of the embryonic rudiments are common in all sorts of animals, including human beings. For example, in mammals that have a coarse diet with considerable amounts of cellulose, a part of the large intestine called the caecum forms a large sac in which mixtures of food and enzymes can react for a long period of time; a constricted appendix is absent altogether. But in human beings the caecum is modified into an appendix.

The nictitating membrane of the eyes is another example. All tetrapods (four-footed animals) have a nictitating membrane in the inner corner of the eye. In most vertebrates, this membrane sweeps clear across the eyeball to cleanse it. In birds this membrane

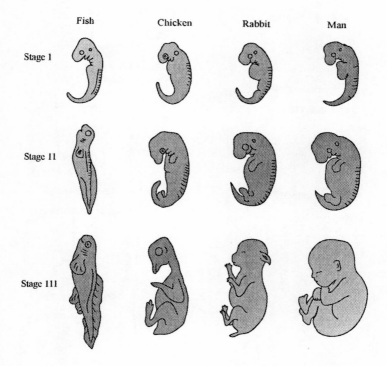

Figure 5-4. EMBRYOLOGICAL EVIDENCE FOR EVOLUTION. Three parallel stages in the development of four vertebrates: fish, chicken, rabbit, and man, show how the early embryonic stages resemble one another. This suggests common ancestry and creation by modification (Islamic concept) or descent with modification (secular concept).

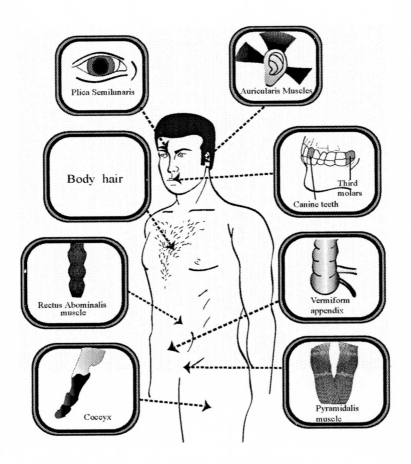

Figure 5-5. FEW OF THE VESTIGIAL STRUCTURES IN HUMAN BEINGS.

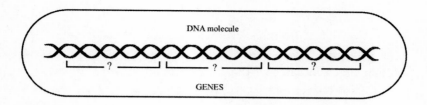

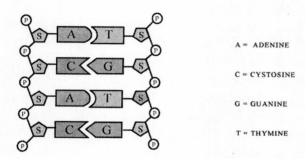

A = ADENINE

C = CYSTOSINE

G = GUANINE

T = THYMINE

Figure 5-6. STRUCTURE OF DNA.

is particularly well developed. In horses and other mammals, it is well developed and fully functional. But in humans and some other mammals, it forms a mere fold at the inner corner of the eye and has no function.

The human external ear muscles present a similar situation. Many mammals move the external ear freely in order to detect sounds efficiently. The complete muscular apparatus for these movements is present in man, but it has no function.[36] (Figure 5-5.)

As previously mentioned, DNA is the cornerstone of all life on earth. It is made out of phosphate, sugar, and four bases—adenine (A), cytosine (C), guanine (G), and thymine (T). (Figure 5-6.)

These bases can be thought of as a four-letter alphabet or code. Just as words in our language differ according to the sequence of letters of the alphabet, the sequence and number of the bases (A, C, G, and T) in the DNA determine the uniqueness of a gene; and combinations and permutations of genes in the chromosome decide the nature and shape of life forms on earth.

The study of chromosomes quite convincingly points out that all organisms, from bacterium to human, are made out of the same substance. Molecularly they differ only in relation to the number and sequence of the four bases in each gene and the combinations of genes in each chromosome, as well as the number of chromosomes in each cell. This unity in diversity suggests the genetic continuity and common ancestry of all organisms.

The evidence of evolution presented above is a fraction of examples too numerous to include here. Unquestionably we can conclude that biologists have posited various theories to explain (1) the phenomenon called life, with all its peculiarities and varieties; (2) the closeness and remoteness in time of various life forms and the interrelationships between them; (3) the geological appearance of less complex life forms during the early stages of the earth as well as the appearance of graduated complexity in the organisms during the long history of the earth; and (4) the interrelationship between organisms and the environments in which they live.

Universally, biologists do not agree on all postulates, but the overwhelming majority agrees that the different forms of life did not originate simultaneously. They also agree that in the early biography of life, organisms were much less complex, but over a period of 3.5 billion years on the earth, increasingly complex life forms emerged. Similarly, almost all biologists concur that organisms can be traced to a single family, based upon the overwhelming evidence of later organisms having evolved from their earlier ancestors.

Evolutionists readily admit that details of such origination and the mechanisms of the transformation of simple organisms into multitudes of species is still a matter of debate. Unfortunately, American Christian evangelical fundamentalist organizations such as the Institute for Creation Research present a distorted version of such admissions, claiming that the evolutionary puzzles yet to be solved somehow negate the facts that scientists do know. Sadly, some contemporary Muslims also are drawn into the ICR's distorted presentation of scientists' genuine disagreements among themselves.

CHAPTER 6

Pre-Darwinian Muslims and
the Theory of Evolution

Muslims had long recognized that humans have an ancestral link to apes, and they taught this fact in schools which, at the time, were extensions of mosques. Nonetheless, the Judeo-Christian peoples in the West were shocked when they first read Charles Darwin's *The Origin of Species* (1859). Their immediate response was generally similar to that of the wife of the Bishop of Worcester who, after hearing that humans evolved from monkeys, was reported to have exclaimed: "Descended from apes! My dear, let us hope it is not so; but if it is, that it does not become generally known."[1]

Darwin's assertions in *The Origin of Species* had once been widely known, however; his book was only a confirmation of centuries-old Muslim knowledge, supported by data he had collected in his travels.

John William Draper (1812-1883), a prominent scientist, evolutionist, and professor of chemistry at New York University, was a contemporary of Charles Darwin. In 1838 he made the first photographic portrait of the moon through a chemical process.[2] Among his many works are his *History of the Intellectual Development of Europe* and *The History of the Conflict between Religion and Science.*

Six months after the publication of Darwin's book, Professor Draper presented a paper at the meeting for the British Association for

Advancement of Science entitled "The Intellectual Development of Europe Considered with Respect to the Views of Mr. Darwin." During the discussion of this paper, Bishop Wilberforce of Oxford contemptuously inquired of Thomas Huxley, an eminent scientist and advocate of the theory of evolution, whether Huxley claimed his descent from monkeys "through his grandfather or grandmother."[3 & 4]

Draper pointed out in one of his later treatises that the Fathers of the Church insisted upon a more recent origin of Creation. He explained such insistence as a justifiable response to claims that God had neglected the human race as a whole in favor of "the few who were living in the closing ages of the world" for salvation. Draper explained that Christian teaching of the story of the perfect world of Adam is a necessary premise for the fall of humanity, which in turn was a prerequisite for "the plan of salvation." In Draper's view:

> [Christian] (t)heological authorities were therefore constrained to look with disfavor on any attempt to carry back the origin of the earth to an epoch indefinitely remote, and on the *Mohammedan theory of evolution* [italics mine] which declared that human beings developed over a long period of time from lower forms of life to their present condition.[5]

Today we know that Darwin's theory of evolution is actually the Muslim theory of evolution. One finds it difficult to believe that European scientists, who were neighbors to the Muslims, were unaware of their theory of evolution when an American scientist, a contemporary of Darwin, knew that the theory originated among the Muslims. Is it possible that Europeans, who followed Roger Bacon's advice to "learn Arabic and Arabic science for progress,"[6] did not know about the Muslim theory of evolution?

In our own time, there is a sharp rejection of evolutionary studies. The free press in the West, especially that of the United States, avoids the mention of the word "evolution" in some public school textbooks, fearing condemnation from some Western religious establishments. For example, *Man: A Course of Study*

(*MACOS*), begun in 1963 by the National Science Foundation, was finally published in 1970 by the Education Development Center as an introduction to evolution for elementary school students. "No commercial publisher would touch the project because religious groups would not endorse the teaching of this type of material," until a small foundation agreed to publish it. By 1974 many school districts in forty-seven states had adopted *MACOS*, but in 1975, when organized opposition began to assert itself, this sales rate plummeted seventy percent. "The National Science Foundation, which had provided $4.8 million to develop MACOS, suddenly was attacked in Congress. The House of Representatives passed the Bauman Amendment in 1975, giving Congress direct supervision and veto power over every single NSF research grant, [and] Nineteen eighty-one saw a drastic cut in federal support for social science research, and science education was virtually eliminated from the federal budget."[7]

Contrary to the current opposition to teaching evolution in American public schools, centuries before Darwin the doctrine of the gradual development of life forms ending in mankind was part of the curriculum in Muslim schools. Draper discredited the Western myth that Lamarck and Darwin were the originators of the theory of evolution and declared that the Muslim theory was more advanced then that of Darwin.

> Sometimes, not without surprise, we meet with ideas with which we flatter ourselves with having originated in our own times. *Thus our modern doctrine of evolution and development were taught in their* [Muslim] *schools* [italics mine]. In fact they carried them much farther than we are disposed to do, extending them even to inorganic or mineral things.[8]

Most Western evolutionists agree that Darwin's grandfather, Erasmus Darwin (1731-1802), influenced his grandson's interest in evolution.[9] Little is written, however, about who influenced their thoughts on the subject.

Erasmus Darwin was a philosopher, a botanist, a poet, and a physician, and the founding member of the philosophical society called "the Lunatics."[10] In Erasmus's day and long before, almost all philosophical and scientific books were translations of or based on Arabic books of Muslim scientists and philosophers.[11] I believe, therefore, that the Darwins learned evolutionary biology from Muslim scientists, a contention supported by the history of medical education in Europe.

Professor Draper describes the state of Islamic and European medicine before the Enlightenment of the West as follows:

> Saracens commenced the application of chemistry, both to the theory and practice of medicine, in the explanation of the functions of the human body and in the cure of diseases. Nor was their surgery behind their medicine . . . . How different was all this from the state of things in Europe: the Christian peasants, fever-stricken or overtaken by accident, hied to the nearest saint-shrine and expected a miracle; the Spanish Moor relied on the prescription or lancet of his physician, or the bandage and the knife of his surgeon.[12]

Will Durant, the American historian, informs us that the Manual for Oculists, written by the great Muslim occultist Ali ibn-Isa, was "used as text in Europe till the eighteenth century." Durant also asserts the importance of the works of Abu Bekr Muhammad al-Razi (844-926), better known in Europe as Rhazes, who was one of many Muslim scientists and healers. According to Durant, Al-Razi's twenty-volume book, *Kitab al-Hawi* (The Comprehensive Book), which covered all branches of medicine and was translated into Latin, was "probably a highly respected and frequently used medical textbook in the white world for several centuries" and was one of nine books used at the University of Paris in 1395.

Another famous figure whose philosophical and scientific contributions are illuminated by Durant is Abu Ali al-Husein ibn

Sina (Avicenna) (980-1037), whose books were encyclopedias of knowledge that included studies in mathematics, physics, physiology, hygiene, therapy, pharmacology, philosophy, metaphysics, theology, economics, politics, and music. His books were taught as main texts in "the universities of Montpellier and Louvain till the middle of the seventeenth century."[13]

Two other Muslim physicians who influenced Europe and European medicine were Abu Bekr ibn Tufail (Abubacer) (1107-1185) and his student, Abu al-Walid Muhammad ibn Rushd (Averroes) (1126-1298). Ibn Rushd wrote an encyclopedia of medicine (*Kitab al-Kulliyat fi-l-tibb*) that was translated into Latin and taught in Christian universities.

All Muslim physicians were evolutionists, and Western historians have acknowledged the fact that books of medicine written by physicians of the Golden Age of Islam served as the standard textbooks used in all medical schools in Europe until the eighteenth century. Therefore, Charles Darwin's (1809-1882), grandfather and father, Erasmus and Robert Darwin, both physicians, were undeniably influenced by the above-mentioned textbooks.

Moreover, the first Latin translation of Abu Bakr ibn Tufail's *The Story of Hai bin Yaqzan (The Journey of the Soul)* by Edward Pocock, Jr., was published in Oxford in 1671. Several editions of this work appeared in the years from 1671 to 1700. Then, in 1708, Simon Okley published the first English translation and Dutch, German, French translations were made in eighteenth and nineteenth century.[14] The publication of many editions and different translations of this book in England and other parts of Europe suggest that it was a very popular book; the probability is great, therefore, that Charles Darwin, his father, and his grandfather, all read it. *The Story of Hai bin Yaqzan* is the story based on evolution. Erasmus Darwin's book *Zoonomia* [Laws of Organic life], which profoundly influenced Charles Darwin, was published in 1794. In *Zoonomia*, Erasmus Darwin described, "millions of ages before the commencement of mankind . . . all warm-blooded animals have arisen from one living filament".[15] as Abu Bakr ibn Tufail and other Muslims told before Erasmus.

Clearly, Erasmus and Robert Darwin most likely learned about Muslim theory of evolution from the translations of Muslim books. A slight chance exists that these erudite men did not know that the authors of their textbooks of medicine and philosophical essays were evolutionists. In that case, such ignorance can only be explained as an unlikely historical oddity. If, as is far more likely, they did learn about evolution from Muslim scholars, their failure to credit their sources appears to be an intentional plan to obscure the pioneering Islamic contributions to the study of evolution. The evidence presented above makes the second premise more likely.

Muslims nowadays, like many non-Muslims in the West, resist the idea of evolution. Curiously these are the same Muslims who boast about the scientific contributions made by their ancestors. They gleefully ask others whether they have heard about Al-Biruni, who, like Bacon, wrote in his *Vestiges of the Past (Athar-ul-Baqiya)*: "We must clear our minds . . . from all causes that blind people to the truth—old custom, party, spirit, personal rivalry or passion, the desire for influence."[16] Muslims never fail to remind us that Ibn Sina's *Canon of Medicine (Qanun-fi-l-Tibb)* was the chief textbook of medicine in European medical schools until the seventeenth century[17] and continue to praise al-Haitham's contributions to science. Yet, if they were told that these great scientists, of whom they are rightfully so proud, were evolutionists, they would be amazed and annoyed.

Most Muslims in the world believe that Adam and Eve were created in Paradise. They have been indoctrinated with the Judeo-Christian belief that God created Adam, then created Eve from Adam's rib. Jews and Christians who accepted Islam early in its history imported this theological doctrine into Islam. We shall discuss later how this Judeo-Christian story of the *ex nihilo* creation of Adam and subsequent creation of Eve was infused into Islamic faith. First, however, a review of early Muslim philosophers' thought will show many of them were evolutionists.

Ibn Khaldun, the most famous Muslim historiographer and
social scientist, who wrote his *Muqaddimah* [An Introduction to
History], states:

> One should then look at the world of creation. It started
> out from the minerals and progressed, in an ingenious,
> gradual manner to plants and animals. The last stage of
> minerals is connected with the first stage of plants, such
> as herbs, and seedless plants. The last stage of plants
> such as palms and vines is connected with the first stage
> of animals, such as snails and shellfish which have only
> the power to touch. The word 'connection' with regard
> to these created things means that the last stage of each
> group is fully prepared to become the first stage of the
> next group. The animal world then widens, its species
> become numerous, and, in a gradual process of creation,
> it finally leads to man, who is able to think and reflect.
> The higher stage of man is reached from the world of
> monkeys, in which both sagacity and perception are
> found, but which has not reached the stage of actual
> reflection and thinking. At this point we come to the
> first stage of man (after the world of monkeys). This is as
> far as our (physical) observation extends.[18]

The reference in the above passage to the "first stage of man"
clearly show that Ibn Khaldun knew that other hominoid species,
more advanced than the monkey but not equal to modern man,
were created before modern man emerged. Moreover, he states
that he arrived at his conclusion from physical observation.

As we have seen in the section on uniformitarianism, Ibn
Khaldun believed that the human races originated as a result of
natural causes. According to the *Muqaddimah,* Ibn Khaldun and
other Muslims believed that a series of transmutations of one
species into another, over a long period of time, resulted in the
gradual evolution of life, from primitive organisms into a bush

with numerous branches. Thus, life forms are not independently created, but are evolutionary products from ancestral species.

Ibn Khaldun concludes his view on the origin of the races as follows: "Physical circumstances and the environment are subject to changes that affect later generations; they do not necessarily remain unchanged."[19] According to him species are not fixed, but subject to changes with a changing environment. He believed that the physical characteristics of organisms are determined by their "essence." He maintained that active nature (*kiyan*) "has the ability to generate substances and change essences"[20] and that earthly existence is a continuum of transformations of essences that occur in stages in a natural order of ascent and descent. He describes the transformation of species into other species as a result of modification of essence (gene) by nature:

> The essences at the end of each particular stage of the worlds are by nature prepared to be transformed into the essence adjacent to them. This is the case with the simple material elements; it is the case with the palms and vines (which constitute) the last stage of plants, in their relation to snails and shellfish, (which constitute) the (lowest) stage of animals. It is also the case with monkeys, creatures combining in themselves cleverness and perception, in their relation to man, the being who has the ability to think and to reflect. The preparedness (for transformation) that exists on either side, at each stage of the worlds, is meant when (we speak about) their connection.[21]

If he were writing today, Ibn Khaldun would replace the word *essences* with the term genes or DNA. He would be saying: "Nature prepares genes (essence) of species to be transformed into genes (essence) of the adjacent species." Similarly, instead of saying "transformed into the next stage by nature," he would say: "Gradual evolution can be explained in terms of small genetic changes (mutation) in the species and by the ordering of this genetic variation by natural selection."

Abu Bakr Muhammed ibn-Arabi (1165-1240), one of foremost interpreters of the Qur'an, was born in Spain. By the time he was twenty-five years old, he was famous throughout the land as a brilliant scholar, a great writer of elegant poetry, and a Sufi of the first order. Between 1193 and 1198, he wrote *Uqlatu'l-Mustawfiz* [Controller of The Wanderer]. When his opinions were severely attacked in the West, he felt his life was becoming intolerable, so in the year 1200 he left Spain for Mecca, after which he traveled extensively until he settled in Damascus. His description of the gradual evolution of species from one origin is similar to that of the other Muslim scholars mentioned in this chapter.

> On they rolled to perfection: Thus the meaner world was born. Mineral passed to vegetable life, out of which animal life was born.[22]

> Then creation continued on earth, minerals, then vegetations, then animals, and then man. God made the last of every one of these kinds. The last of the minerals and the first of the vegetations is the truffle. The last of the vegetations and the first of the animals is the date-palm. The last of the animals and the first of mankind is the monkey.[23]

> Since the perfect man is in the perfect form, he deserves the vicegerentship and deputyship of God in the Universe. Here we shall explain the evolution of this vicegerent, his position and form as they are. We do not mean Man only as animal, but, on the other hand, as Man and vicegerent. On account of his human quality and vicegerentship man deserves his perfect form. Every man is not a vicegerent. In our opinion, the animal man is not a vicegerent . . . . This is the intended Perfect Man. The others are animal men. The relation of the animal

man to the Perfect Man is that of the ape to the animal man.[23] As for the animal man, he is not a man essentially. His case is like that of animals. But he is distinguished from another through differentiae peculiar to every one of the animals.[25]

The goal of all this was man coming in perfect form. When the field was thus prepared, Man came in the nicest form.[26]

When God desired the perfection of human evolution, He collected and bestowed on Man all realities of the Universe and illuminated him with all His names.[27]

When this comprehensive name became capable of two aspects by itself, it became fit for vicegerentship and organization and gradation of the Universe. If Man does not reach the stage of perfection, he is an animal whose appearance resembles the external appearance of man. Here we are concerned with Perfect Man. The first of human species whom God made was the Perfect Man. He was Adam (peace be upon him). Thus God demonstrated the stages of perfection for the species. He who attains to it is the man who attains perfection, and he who goes down from that stage is one who possesses the human quality in proportion to where he is.[28]

Ibn Arabi explains the concept of the "perfect man" as being God's intended vicegerent on earth. In order for a human being to become God's deputy on earth, he had to evolve into a perfect form. Ibn Arabi makes a distinction between a "perfect man" and an inferior form, which he calls "animal man." The latter is "not a man" because he exhibits qualities peculiar to animals. His relationship to a human being is like "that of the ape to the animal man." In Ibn Arabi's view, if man does not reach the stage of perfection, he is an animal whose appearance resembles the external

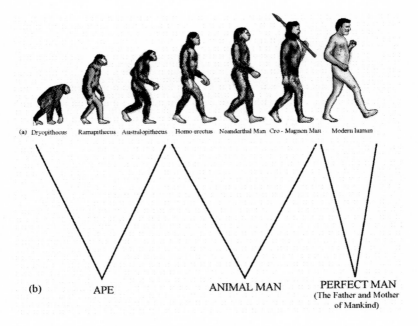

(a) Dryopithecus Ramapithecus Australopithecus Homo erectus Neanderthal Man Cro - Magnon Man Modern human

(b)   APE   ANIMAL MAN   PERFECT MAN
                                                      (The Father and Mother
                                                          of Mankind)

Figure 6-1. (a) An artistic reconstruction of the evolution of man and the classification of species based on fossil evidence. (b) The pre-Darwin Muslim classification of man and his ancestors.

appearance of man. Thus, it was God's intention for human beings to evolve into their perfect form[s] so they would be suitable vicegerents of the Universe.

The above ideas from Ibn Arabi demonstrate that the Muslims knew about the existence of hominoid species after the advent of the ape and before the emergence of modern Homo sapiens. Not only did Ibn Arabi classify the hominid as an "animal man" but made the distinction between that species and the human being he called "perfect man."

Gould asked "why the Lord saw fit to make so many kinds of hominids, and why some of his later productions, Homo-erectus in particular, look so much more human than the earlier models?"[29] Ibn Arabi anticipated these questions long before Gould or Darwin. Fossils of the animal man to which Ibn Arabi referred were only recently unearthed and classified as Australopithecus, Homo erectus, Homo ergaster, Neanderthal man, and Cro-Magnon man. An artist's reconstruction of human evolution based on modern fossil findings (Figure 6-1) fits well with Ibn Arabi's description of the origin of humans from earlier species.

It is possible that Ibn Arabi and his contemporaries arrived at the conclusion that man evolved from "animal man" to human not by pure speculation, but by scientific methods. For example, Al-Biruni came to the conclusion that there were hominids of different statures by studying the size of their bones, which were buried in the caves of the mountains of Median where they lived. Al-Biruni describes these bones as those of large races of men "with bones as large as camel-bones and even larger."[30]

Predating Ibn Arabi, in the tenth century Ikhwan al-Safa (the Brethren of Purity) wrote about the gradual development of life. They maintained that species developed successively from minerals to plants to animals and finally into humans, that imperfect species preceded the more perfect ones, and that God created them for humankind's sake. They pointed out that "by the grace of God" nature provided each species with the necessary organs to insure its survival and secure the completion of its perfect form. These

conclusions are derived from the following quotations from the
Rasa'il of Ikhwan al-safa (The Epistles of Ikhwan al-Safa).

> The first stage of the plant kingdom is connected with
> the last stage of minerals and the highest stage of the
> plant kingdom with the first stage of animal . . . . Also we
> would like to show in this letter that the highest stage of
> the animal kingdom is connected with the first stage of
> the human.[31] Be it known to you! The imperfect animals
> preceded the most perfect animals in time and in the
> process of creation.[32] Be it known to you, brothers! The
> animals were created for man's sake. (And) everything
> that is created for the sake of something else will precede
> the beneficiary.[33] By the grace of God's wisdom and care,
> animals were bestowed with organs, joints, vessels,
> nerves, membranes, and chambers according to their
> needs for benefiting themselves or to avoid injury, so
> that it can survive and will be completed and perfected
> to reach the highest stage.[34]

The celebrated mystic poet Jalaluddin Rumi (1207-1273) was
born in Balkh, Afghanistan. His father was the son of Bahu'u'ddin
Walad, a well-known professor of theology and a student of Ibn
Arabi. Rumi's celebrated philosophical treatise, *Masnawi-ye Ma'nawi*
(or *Mathavi* in Persian, Rhyming Couplets of Deep Spiritual
Meaning), has been linked to the writings of Ibn Arabi. Therefore,
Ibn Arabi's thoughts may have influenced Rumi's views on the
creation of man.[35] In his treatise Rumi echoed the same perspectives
on the origin of man and life as that preached by other Muslim
scholars of his time and before. In the *Masnavi,* he wrote:

> Hundreds of thousands of years I was flying (to and fro)
> involuntarily, the notes in the air.[36]

> He came first to the inorganic realm and from there
> stepped over to the vegetable kingdom. Living long as a
> plant, he has no memory of his struggles in the organic

realm. Similarly rising from the plant to the animal life he forgets his plant life, retaining only an attraction for it which he feels especially in the spring, ignorant of the secret and cause of his attraction like the infant at the breast who knows not why he is attracted to the mother . . . . Then the creator draws him from animality to humanity. So he went from realm to realm until he became rational, wise, and strong. As he has forgotten his former types of reason (every stage being governed by a particular type of reason) so he shall pass beyond his present reason. When he gets rid of this coveted intellect, he shall see a thousand other types of reason.[37]

I died from the inorganic realm and became a plant; then I died from plant life and became an animal. Dying from animality I became a man, so why should I be afraid of anything less through another death? In the next step I shall die from humanity to develop wings like angels. Then again I shall sacrifice my angelic self and become that which cannot enter imagination. Then I become non-existent when the divine organ strikes the note, "We are to return unto Him".[38]

According to Rumi, the so-called evolution and natural selection were acts of God, who "draws him from animality to humanity." Further, Rumi envisions creation as a journey through which a human being necessarily experiences the passage from a physical to a metaphysical existence, thereby returning to God, who is the ultimate source of creation.

Islam's greatest geologist, Al-Biruni (973-1048), believed in the gradual development of man and other organisms, and in natural selection through the struggle for the survival of the fittest. He wrote about his beliefs regarding the progressive development of mankind in his book, Kitab al-Jamahir.

Man reached his maximum degree of perfection compared to other animals below him (in the evolutionary ladder). He ascended to the present state from other kinds of beings such as dog-like, bear-like, ape-like, etc. Then finally he became man.[39]

Abu Bakr Ibn Tufail (1095-1138) (Latin name: Abu Bazer), a poet and an eminent physician, was born in Guuadix, Spain and died in Morocco. He was known to have influenced Jewish and Christian thinkers and was the teacher of the great philosopher Ibn Rush (Latin name: Averroes). Although he wrote treatises on medicine and philosophy, Ibn Tufail is remembered for his famous work, *The Journey of the Soul (The Story of Hai bin Yaqzan)*.[40] The book is a fictional biography that deals with the idea of evolution. Simon Ockley undertook the English translation in 1708, about a century before Darwin was born. H. F. Osborn summarizes the story as follows:

> There happens to be under the equator an island, where Man comes into the world without father and mother; by spontaneous generation he arises, directly in the form of boy, from the earth, while the spirit, which, like the sunshine, emanated from God, unites with the body, growing out of a soft, unformed mass. Without any intelligent surroundings, and without education, this 'Nature-man' through simple observation of the outer world, and through the combination of various appearances, rises to the knowledge of the world and of the Godhead. First he perceives the individuals, then he recognizes the various species as independent forms; but as he compares the varieties and species with each other, he comes to the conclusion that they all sprung from a single animal spirit, and at the same time that the entire animal race forms a single whole. He makes the same discovery among the plants; finally he sees the animal and plant forms in their unity, and discovers that among all their differences they have sensitiveness and feeling in common; from that he concludes that animals and plants are one and the same."[41]

Osborn points out that the boy compares various species of plants and animals and concludes that they sprang from one common origin. Those who wish to preserve science from the corruption of religion echo the same sentiments. Robert J. Shadewald, the former editor of the *National Center for Science Education Report,* asserts that, "conventional science offers the same simple answer—all animals have descended from a common ancestry."[42]

Eight hundred years before Darwin, Ibn Miskawayah (930-1030), a Persian scientist, philosopher, and historian, was interested in alchemy and poetry and left a legacy of twenty works that served as models for later generations of Islamic thinkers.[43]

Allamah Mohammad Iqbal, a renowned twentieth-century Muslim poet from the Indian subcontinent, summarized Ibn Miskawayah's views on the origin of life. Iqbal explains that this Muslim scientist found that "plant life, in its lowest forms, does not "need any seed for birth or growth." The difference between this kind of plant life and minerals is "only in some little power of movement." This power of movement "grows in higher forms, and reveals itself further in that the plant spreads out its branches, and perpetuates its species by means of its seeds." Ibn Miskawayah believed that the last stage of plant development is the vine and the date palm, "where a clear sex differentiation appears." He maintains that the beginning stage of animality starts when plants are released from earth-rootedness, start to develop "the germ of conscious movement," and begin to acquire movement through the development of the senses of touch and sight, followed by the other senses. He believed: "Animality reaches its perfection in the horse among the quadrupeds and the falcon in birds, and finally arrives at the frontier of humanity in the ape, which is just a degree below man in the scale of evolution." The final stages of the evolution of animals bring psychological changes with the growing power of discrimination and spirituality until humanity passes from barbarism to civilization.[44] Ibn Miskawayah believed and taught that plants, animals, and human beings belong to the same family, which originated from simple life forms that gradually spread all over Mother Earth.

Abu Hassan al-Masudi (d. 957), a Muslim from Baghdad, was "the Pliny as well as Herodotus of the Muslim world."[45] He traveled extensively and gathered his gleanings into a thirty-volume encyclopedia in which he "surveyed omnivorously the geography, biology, history, customs, religion, science, philosophy, and literature of all lands from China to France." Al-Masudi summed up his scientific ideas in *A Book of Information* in which "he suggested an evolution from mineral to plant, from plant to animal, and animal to man."[46] Similarly, Uthman Amr al-Jahiz (776-868), a theologian and litterateur of Ethiopian origin, asserted that evolution occurred gradually, progressing from plant to animal to man.[47] Al-Jahiz's major works are *Book of Animals (Kitab al-Hayawan)* and *Elegance of Expression and Clarity of Expression (Kitab al-Bayan we al-Tabyin)*.

Muhammed al-Haitham (965-1039), a Muslim scientist from Egypt, was known to medieval Europe as Alhazen. Durant acquaints us with al-Haitham, "without whom we might have never heard of Bacon."[48] He is famous for his *Book of Optics (Kitab al-Manazir)*. He "studied the refraction of light through transparent media like air and water, and came so close to discovering the magnifying lens that Roger Bacon, Witelo, and other Europeans three centuries later based upon his work their own advances towards the microscope and the telescope."[49]

Al-Haitham defended the doctrine of evolution. We do not have his original work, but we have two secondary sources. In one of them Ameer Ali, a historian, tells us that Muslim philosophers, including al-Haitham, accepted the theory of evolution. He summarizes their views on the subjects: "In the region of existing matter, the mineral kingdom, then animal kingdom, and finally the human being. By his body he belongs to the material world, but by his soul he appertains to spiritual or immaterial."[50] William Draper confirms al-Haitham's belief in evolution and his strong belief that "man, in his progress, [passed] through a definite succession of states" before he became the present day man.[51]

Ibrahim al-Nazzam, who lived about a thousand years before Darwin, was described as a "brilliant Muslim theologian of classical Islam."[52] He said that creation took place only once with infinite possibilities and potentialities. He taught that all elements that existed in the past or will exist in the future "were latent or potentially present in the stuff originally created—minerals, plants, animals, and men" and that they are "only the gradual realization of those latent potentialities."[53]

Although early Muslim philosophers and scientists believed that the environment modified organisms and human beings, they held that changes in life forms were God's way to create and destroy. Hence, Ibn Khaldun concluded the paragraph on the origin of the human races (refer to chapter 3) with a quote from the Qur'an: "This is how God proceeds with His servants—And verily, you will not be able to change God's ways."[54]

NATURAL SELECTION

Western scholars and scientists insist that Darwin formulated the theory of natural selection. Gould points out that within a decade, Darwin convinced the scientific world that evolution had occurred, but his theory of natural selection never achieved much popularity during his lifetime. It did not prevail until the 1940s.[55]

On the contrary, Muslims have long described natural selection working within the process of creation. Darwin borrowed the idea of artificial selection from horticulturists and farmers to explain natural selection. He pointed out that "when a race of plants is . . . well established, the seed-raisers do not pick out the best plants, but merely go over their seed-beds, and pull up the 'rogues,' as they call the plants that deviate from the proper standard." Darwin asserts that with the help of variations provided by Nature, humans apply the principles of selection to adapt organic beings to their use, but asserts that "natural Selection . . . is a power incessantly ready for action, and is immeasurably superior to man's feeble efforts."[56]

According to historical records, almost ten centuries before Darwin, al-Biruni explained natural selection. He believed that God delegated active nature (*kiyan*) to perform its assigned duties in the universe. He explained natural selection in almost the same words as did Darwin centuries later. He also saw examples of selection in the methods of horticulturists, as well as in the natural behavior of bees. Al-Biruni wrote in *Fi Tahqiq Ma Li'l-Hind* (*Al-Biruni's India*):

> The agriculturist selects his corn, letting grow as much as he requires, and tearing out the remainder. The forester leaves those branches, which he perceives to be excellent, whilst he cuts away all others. The bees kill those of their kind who only eat, but do not work in their beehive. Nature proceeds in a similar way.[57]

The following quotes from al-Biruni support the idea that long before Malthus and Darwin, he knew about the disparity between reproduction and survival. He describes here speciation and natural selection as result of structural advantages in some species of plants and animals. Al-Biruni wrote:

> The life of the world depends upon sowing and procreating. Both processes increase in the course of time, and this increase is unlimited, while the world is limited. *When a class of plants or animals does not increase any more in its structure, and its peculiar kind is established as species of its own* [italics mine], when each individual of it does not simply come into existence once and perish, but procreates a being like itself or several together, and not only once but several times, then this will, as single species of plants or animals, occupy the earth and spread itself and its kind over as much territory as it can find.[58]

Notice the similarity between the above quote from al-Biruni and the following passage from Darwin:

> A struggle for existence inevitably follows from the high
> rate at which all organic beings tend to increase [and]
> every organic being is striving to increase in a geometrical
> ratio; that each at some period of its life, during some
> season of the year, during each generation or at intervals,
> has to struggle for life and suffer great destruction. When
> we reflect on this struggle, we may console ourselves
> with full belief, that the war of nature is incessant . . . the
> vigorous, the healthy, and the happy survive and multiply
> [and] that any variation in the least degree injurious
> would be rigidly destroyed. This preservation of
> favourable individual differences and variations, and
> the destruction of those, which are injurious, I have
> called Natural Selection or the Survival of the Fittest.[59]

There is little difference in the views of al-Biruni and Darwin.
Both agree that organic beings have unlimited ability to procreate,
but that the world is limited. And both concur that individual
variations in the structure of organic beings determine the survival
of the peculiar kind among the species. Although Darwin labeled
the natural phenomenon as "natural selection," al-Biruni closely
described it when he used phrases such as "nature proceeds in a
similar way," "the agriculturist selects" or "the forester preserves . . .
while he cuts away all others."

Materialists argue that the imperfections in nature suggest the
absence of God in the evolution of life. Al-Biruni found that nature
allows for the existence of evolutionary imperfections, citing the
example of animals with imperfect limbs. He wrote:

> [If] Nature, whose task it is to preserve the species as
> they are, finds some superfluous substance . . . she forms
> [it] into some shape instead of throwing it away; when
> Nature does not find the substance by which to complete
> the form of that animal in conformity with the structure
> of the species to which it belongs . . . she forms the animal
> in such a shape, so that the defect is made to lose its

obnoxious character, and she gives it vital power as much as possible.[60]

Furthermore, like other Muslim scholars of his time, al-Biruni believed that while these imperfections may appear random to human beings, they are carried out by nature as a part of the "official function" designed for it by Allah.

> Frequently . . . you find in the functions (actions) of Nature which it is her office to fulfill, some fault (some irregularity), but this only serves to show that the Creator who had designed something deviating from the general tenor of things is indefinitely sublime, beyond everything which we poor sinners may conceive and predicate Him.[61]

Regarding natural selection, Jalaluddin Rumi details the natural struggle for survival as the essence of life and a testimony of God's Will. He points out that the natural elements, such as air, earth, water, and fire, exist to serve God and that life is a universal struggle.

> The Universe, when you look at it, closely presents a universal struggle—atom struggling with atom like faith against infidelity. The struggle in action is the objective form of the principle of opposition, which has its basis in their inner nature. There is war in words and war in deeds and war in nature; between the parts of the Universe there is a terrible war. This war is the very constitution of the Universe; look into the elements and you understand it. Creation is based on opposition; therefore, every creature became warlike to get some benefit and avoid some injury. This struggle is not the phenomenon only of outward nature; even thy own self is a battlefield of one mental state opposing the other. The essence of soul transcends these oppositions; its nature is not (contradictory) like this; it is divine. Only

opposites destroy opposites. Where there is no opposite
there is eternal life.[62]

Because without need Almighty God does not give
anything to anyone; if there were no necessity, the seven
heavens would not have stepped out of non-existence;
the sun and the moon and the stars could not have
come into existence without necessity; so necessity is
the cause of all existence, and according to his necessity
man is endowed with organs. Therefore, O needy one,
Increase your need so that God's beneficence may be
moved (to bestow new instruments of life on me).[63]

Unlike modern hardcore evolutionists, such as Stephen Jay
Gould and others, who believe in the absolute mastery of nature,
al-Biruni and other Muslims see nature not as an absolute free
agent, but as a force that operates in compliance with divine
providence and whose actions are sometimes difficult to
understand due to the limitations of the human mind. However,
without God, the evolver, man probably would not have come
into existence. Gould accepts that if we were to rewind the videotape
and replay the movie, the appearance of humankind on the earth
is unlikely.[64] This is probably one of the reasons that Muslim
philosophers and scientists believed that without divine guidance
for the imperfect nature, man would not have evolved. Muslims
accept that all created beings are imperfect in structure and wisdom.
The only one who is perfect is God, and so He is called *Al-Hakim*
(the Perfect in Wisdom).

Based on the pre-Darwin Muslim works that have been cited,
we can conclude that Muslim scholars viewed life as a bush and
all life forms as either twigs or branches of that bush. They believed
that humans are related to apes as apes are related to their
predecessors. Al-Biruni's and Rumi's explanations of natural
selection were rendered almost verbatim by Darwin centuries later.
Yet, Western historians scarcely mention Muslim names when they
record the history of evolution.

The Western claim that Aristotle is the originator of the ladder of nature is an inference that cannot be corroborated because "not a single original word that Aristotle wrote exists today."[65] By contrast, we have detailed descriptions of the ladder of nature in Muslim works. It is unlikely that Muslims derived their evolutionary theories from Judeo-Christian teachings or from Darwin and Spencer. The abundant evidence presented here demonstrates that Muslims are the originators of the theory of evolution, and William Draper is correct when he calls it the Muslim Theory of Evolution. The only difference one can cite between the Muslim scholars and Darwin is that Muslims believed that the existence of the ladder of nature was the result of divine will and providence.

Why did Muslim scholars develop the evolutionary paradigm of the origin of life rather than accept the existing belief of their time? Because they derived their thought from their divine book, the Qur'an, in addition to their independent scientific observations. These inquisitive but believing Muslims took seriously the following commands in the Qur'an as a methodology for a correct understanding of the world:

> On the earth are the signs for those who have sure faith
> (in the meaningfulness of all things), as also (there are
> signs) in your own self: will you not, then, observe? [66]
> (Qur'an 51:20-21)

The Qur'an further commands us to search and learn from the earth the process of the creation of life, man, and the universe (Qur'an 29:20).[67] Early Muslims considered the universe as another book of God, and in trying to decipher the laws of nature, they were also following God's commands.

The above verses suggest that the universe, with all its physical, chemical, biological, and psychological laws, can be comprehended. They also suggest that every Muslim should understand God's method of creation. What is the situation with Muslims of today? It is a sad story! Instead of intellectual pursuit enjoined by God's guidance, intellectual oppression is the rule in most Muslim countries. The Prophet said, "One learned man is harder of the devil than a thousand ignorant worshippers;" and "The ink of the scholar is holier than the blood of a martyr."[68] Unfortunately, the sacred ink has dried up and unholy blood is dripping from Muslim pens in many part of the world, including at the site of the World Trade Center. I believe there is an urgent need for Muslims to reacquaint themselves with the meanings of the Qur'an. Therefore, let us next review the Qur'an to find out what it says about the origin of man.

CHAPTER 7

The Qur'an and the Origin of Man

The Qur'an does not have a CHAPTER on the genesis of human life, as do the Jewish and Christian scriptures. Christians and Jews believe that God created human beings in His image and created Eve from Adam's rib. Most contemporary Muslims also believe that God created Eve from Adam's rib, although not a single verse in the Qur'an supports this belief. What, then, does the Qur'an say about the creation of human beings?

The Qur'an is a book of guidance that asks its believers to investigate and understand nature. It does not spoon-feed them with knowledge, but advises them to observe and reflect on nature. An example of this method of teaching can be observed in the following verses:

> Truly in the creation of heavens and the earth, and in the alternations of the night and day are signs for those who have acumen, who utter (the name of) God, standing, sitting, and on their sides, and ponder over the creation of the heavens and the earth, (saying), "Our Lord, You have not created these in futility. Glory be to You: guard us then from the torment of the fire." (Qur'an: 3:190-191)[1]

> On the earth are signs for those who have sure faith (in the meaningfulness of all things), as also (there are signs) in your own self: will ye not, then, observe? (Qur'an: 51:20-21)[2]

The verses instruct that contemplating God's creation, pondering the open book of the universe, observing God's creative hand moving this universe, turning the pages of the book, and learning from human anatomy, all evoke true worship and remembrance of God. This verse and other verses encourage Muslims to examine the world with a scientist's eyes to understand the products and process of creation. Thus, our perception of the process of creation is enriched as we continue to study all pertinent verses of the Qur'an and to explore the universe as new technological means become available to us.

Verses about the origin of life and man are scattered throughout the Qur'an; we have collected the key verses in this chapter in order to develop a clear picture of the meanings they convey. A casual reading of the verses may give us the impression that they contradict each other, but apparent contradictions evaporate when we interpret the meanings of all the verses collectively rather than the meaning of each single verse.

More importantly, our conclusions should be consistent with our overall reading of the Qur'an. Muslims accept the Qur'anic assertion that its internal consistency is proof of its divine origin:

> Do they not consider the Qur'an (with care)? Had it
> been from other than God, they would have found
> therein much discrepancy. (Qur'an: 4: 82)[3]

Thus, Muslims generally contend with verses that seem contradictory in one of three ways: They conclude either that their understanding is wrong, or that the totality of knowledge at a given time and place is not sufficient for them to understand the purpose and message of the verses, or verse is allegorical.

The Qur'an informs us that all things are created by God in accordance with His grand design, that God is manifested through His creations, and that His attributes are the link between human beings and their comprehension of the Divine.[4] Out of the titles and ninety-nine attributes ascribed to God in the Qur'an, four grant us insight into His process of creation. Those key attributes are *Rabb* (the Sustainer), *al-Khaliq* (the Creator), al-*Baari* (the Evolver), and *al-Musawwir* (the Bestower of Forms).

The first command that came to the Prophet was:

> Read in the name of thy Sustainer (Rabb), who has created; created man out of a germ-cell. (Qur'an: 96: 1-2)[5]

The general translation of the word *Rabb* as Sustainer does not convey its overall meaning. The noun *Rabb* is derived from the Arabic word *Rububiyat,* the meaning of which cannot be fully rendered in English with one word. Based on his analysis of the works of early Arab lexicographers, Abul Kalam Azad, a Muslim Qur'anic scholar and commentator from the Indian subcontinent, deciphers the meaning of the word as follows: "To develop a thing, stage by stage, in accordance with its inherent aptitude and needs, its different aspects of existence and also in the manner affording the requisite freedom for it to attain its full stature."[6] Likewise, Imam Aul'l-Qasim ar-Raghib (11th century), in his book *Al-Mufradat fi Gharib al-Qur'an (The Vocabulary of The Qur'an),* defines the meaning of the word *Rabb* as follows: "*Rabb* signifies the fostering of a thing in such a manner as to make it attain one condition after another until it reaches its goal of perfection."[7]

The major components of the meaning of the word *Rububiyat* are thus: (a) development of a thing by an external agent; (b) a step-by-step process, not an instant event; and (c) the freedom for the objects "to attain full stature" within the overall creative process. Therefore, *Rabb,* the derived noun from *Rububiyat,* means an evolver. The use of the noun *Rabb* as an attribute of God suggests that God lets organisms evolve, affording them the freedom to attain complete perfection within the limits of His laws of nature.

Another name of God is *al-Khaliq.* The word is derived from the Arabic root verb *khalaqa,* which is used in nearly all the verses pertaining to the creation of human beings and the universe as a whole. To explain how the Qur'an describes this process, we first need to understand the meaning of *khalaqa.* Almost all commentators and translators of the Qur'an, both Muslim and non-Muslim, translate the meaning of this verb as corresponding to the English verb "to create." However, this translation does not convey the original and complete meaning of the verb.

The Arabic-English lexicon compiled in 1883 by a British lexicographer, Edward William Lane, guides non-Arabic speaking students of the Qur'an to understand its meanings. At the request of the Duke of Northumberland, Lane edited his lexicon after twenty-five years of exhaustive study of the Arabic language. He explains that the original Arabic language of Ma'ad, during the time of Mohammad, was a highly complex and difficult linguistic system. As a result of Muslim conquests and the intermingling of Arab culture with that of other nations, the classical Arabic language lost its original character and assumed a simpler version of the original language that became predominant even in Arabia.[8]

Lane's lexicon is the most authoritative for the non-Arabic-speaking students of Islam because he traced the original meanings of the words back to the time of Ma'ad and Prophet Mohammad. His interpretation of Arabic words is based on earlier lexicographers and grammarians, such as Al-Khaleel (author of Eyn, died in 667 AD), Esh-Sheybanee (author of Jeem, Nawadir, and El-Ghareeb el – Musannaf, died in 802 AD), Al-Baydawee (author of Exposition of the Qur'an, 1290 AD), Al-Feiyoomee (1333 AD), al-Firuzabeedi (author of Kamoos, 1330-1421 AD), and Seyyid Murtada-al-Zabeedi (author of Taj-al-Aroos, 1732-1791 AD).[9]

According to Lane, the verb *khalaqa* means "proportioning a thing into another thing" and "to bring a thing into existence according to a certain measure, or proportion, and so as to make it equal to (another thing)."[10] In "proportioning a thing into" an original and previously non-existing thing, it also signifies "the originating, or to bring a thing into existence after it had not been, or the bringing a thing into existence from a state of non-existence."[11]

Hence, there are three components to the meaning of the word *khalaqa*: (a) shaping an original substance or entity into another object ("proportioning a thing to another thing"); (b) the newly formed object or creature must have its own peculiar characteristics so that it can be identified separately from its original source—not like father and son, but like a tree to a canoe or an ape to a human; and (c) the newly formed object or creature with its characteristic features were nonexistent before its original birth ("to bring a thing into existence from a state of non-existence"). In this process the new creature becomes a prototype. Therefore, the classical meaning of the Arabic word khalaqa can be summarized as follows:

> To bring a thing into existence according to a certain measure, or proportion, so as to make it equal to another thing that is not pre-existing.[12]

The above meaning of *khalaqa* is more befitting with the overall reading of the Qur'an because it is based on the true language that was prevalent among the natives of the Arabian Peninsula at the time of the Prophet Muhammad and not upon modern Arabic. Moreover, it does not contradict what we observe in nature, because the meaning of the verb *khalaqa* implies a realization that various life forms were not created simultaneously on the earth, but in stages.

Now that we understand the true meaning of *Rabb* and *khalaqa* and in order to gain a better understanding of the way the Qur'an describes the process of creation, those of us who read English translations of the Qur'an should substitute the above meanings wherever the verbs *create* or fashion appear.

The Qur'an describes God as "the Evolver" (*al-Baari*) and "the Bestower of forms" (*al-Musawwir*), reinforcing the idea of creation as a stage-by-stage process. The verse reads: "He is God, the Creator [*al-Khaliq*], the Evolver [*al-Baari*] and the Bestower of forms [*al-Musawwir*]" (Qur'an: 59:24). [13]

Al-Baari, the Arabic word in the above verse, is derived from the verb *baara*, which means "a thing's becoming clear, or free, of, or from another thing; either by being released [therefrom]" or by "evolution from a previously created matter or state."[14 & 15] So God, the executor of such evolution, is *al-Baari* (the Evolver).[16]

The word *al-Musawwir* is derived from the verb *sawwara*, which means to sculpt a thing and give definite form or color to make things exactly suitable for a certain end or object.[17] Hence God is called *al-Musawwir*, or the "sculptor, of all existing things, who has established them, given to every one of them a special form and a particular manner of being whereby it is distinguished, with their variety and multitude."[18]

We would not be wrong to infer, then, that if God is the evolver, He is also the creator of new life forms from those He had previously created. Moreover, since He is the "Bestower of forms and colors," so His creations function suitably in environments to achieve "certain ends or objects," we could logically infer that, by changing the genetic configuration, He creates new species capable of adapting to the changes in environment over the various ages of the earth. The Qur'an asserts that such changes are the result of the will of God.

In summary, the meanings of *Rabb, al-Khaliq, al-Baari,* and *al-Musawwir* confirm the belief that evolution and creation are not contradictory but mutually complimentary.

We learn from the Qur'an that the creation of the universe and its contents was not an instant event, but a process: "It is He who beginneth the process of creation (khalaq) and repeats it." (Qur'an: 10-4)[19] Yusuf Ali, a well-known Muslim translator, explains, "God's creation is not a simple act, once done and finished with. It is continuous, and there are many stages, not the least important of which is the Hereafter, when the fruits of our life will be achieved."[20] In support of this inference we cite:

> He who has made everything which He has created most good: He *began the creation (khalaq)* of man with (nothing more than) clay, and made his progeny from a quintessence of a fluid despised. (Qur'an: 32:7)[21]

and,

> Verily We created man from a sperm yoked (to the ovum) . . . (Qur'an 76:2)[22]

The creation of Prototype humans, Adam and Eve, began from clay. However, their progenies were created by an "emitted drop of semen" "yoked with ovum." Through scientific evidence, we know that the creation or birth of a human baby involves the evolution of cells through a process of meiosis, during which the nucleus of a cell divides and forms a mature germ cell comprised of sperm and ovum that unite in a continuous process of reproduction.

Although the Qur'an does not render an exact explanation of the above, its verses lead us to infer that the evolution of yoked sperm and ova through all later stages of intrauterine fetal growth to form human progeny does not occur instantaneously, but as a step-by-step process rooted in finite time and space in relation to an earthbound person's time frame.

Similarly, the creation of prototype Homo sapiens began with clay. The phrase in the verse 32:7, "He *began* the creation of man (emphasis added)" indicates that the beginning was not an end in itself. Additional steps necessarily followed to complete the intended task. Other verses also clearly imply a time lapse, for example: "He fashioned (*khalaqa*) you from sounding clay, like unto pottery." (Qur'an: 55:14) [23] The word *fakha'ar*, in Arabic text of the Qur'an, means "baked pottery or a baked vessel of clay."[24] We can read this verse as follows: "Just as clay is molded into shape in stages and baked into pottery, prototype human was also created through successive stages over a period of earthly time." This inference is supported in the following verses: "He fashioned you, and perfected your shapes" (Qur'an: 64:3) [25] and "He created (*khalaqa*) you in successive stages" (Qur'an: 71:14). [26] The latter verse unequivocally states that the creation of mankind was not a magical *ex nihilo* instant event but a step-by-step transformation.

Today, we scientifically explain these stages as a process that expands from nothingness to the Big Bang, from the Big Bang to the earth, from clay to nucleotide, to deoxyribonucleic acid (DNA) to microorganism, from microorganism to marine animals, from marine animal to ape, from ape to Australopithecus to Homo erectus to Homo ergaster. Homo ergaster branched into Neanderthal and

Cro-Magnon man. Cro-Magnon evolved in to Modern man. Modern agnostic biologists describe the "creation by stages" as follows:

> Evolution usually proceeds by "speciation"—the splitting of one lineage from a parental stock—not by slow and steady transformation of these large parental stocks. Repeated episodes of speciation produce a bush.[27]

According to the Qur'an, man is created from components: "Who created (khalaqa) you, then proportioned you into whatever form He willed. He made you out of components [rakkaba]." (Qur'an: 82:7-8). The verb rakkaba in the original Arabic text of this verse means "to create a thing from components; put or set one part of it upon another."[28] Therefore, the verse indicates that the necessary components were created as prerequisites for the creation of man.

Even though the Qur'an states that God created humans in successive stages, it does not describe all the stages of their evolution from clay before they became Homo sapien sapiens (modern man). Sperm and ova are not human. The forty-six chromosomes present in the fertilized human egg are the encapsulated secret codes of a future human being, but we do not use the term man or woman to describe them. They are still a bundle of amino acids, phosphate, and sugar. A thing is named only when it comes into its own form and acquires specific characteristics. The stages through which human beings came into existence have their own names—water, clay, soil, microorganism, marine animals, mammal-like reptiles, apes, Australopithecus, Homo erectus, Homo ergaster, and Cro-Magnon man. However, all those are only landmarks in the journey of the creation of mankind. They took place a long time ago, and memories of those stages do not remain in the human brain, but are stored in the vestigial structures, fossils, chemistry, immunology, and so on.

Those unremembered stages of evolution in the creation of mankind might be what the following verse alludes to:

> Has there come on man a while of time when he was a
> thing unremembered? (Qur'an: 76:1)[29]

The verse suggests that mankind existed in some kind of form that was not recognizable as human. Perhaps this was one among many verses that may have inspired Jalaluddin Rumi (1207-1273), a great Muslim sage, to compose the strophe quoted in the previous chapter, in which he describes the evolution of humanity stage by stage without remembering his earlier stages.[30]

The Qur'an reads:

> He said, "So it shall be." Your Lord said, "Easy it is for Me,
> as I created you before you were nothing." (Qur'an: 19:9)[31]

This verse tells us that mankind and the universe were created from nothingness. The initial void, with the exception of the existence of God, was nothingness—no space or time. When the Qur'an refers to a state of nothingness, it is not saying that human beings were created *ex nihilo* without any connection to other life forms or species. If Muslims do not accept this inference, then they have to admit that chapter 19, verse 9 is contradictory to many other verses in the Qur'an. They need to ask the question: If man was created *ex nihilo,* why does the Qur'an state that man was created from water, clay, soil, and germ-cell, and in successive stages?

Almost all biologists accept the fact that life originated in water. This is what the Qur'an reveals about the aquatic origin of life:

> Do not the unbelievers see that heaven and earth were
> joined together (as one unit of creation), before We
> clove them asunder? We made from water every living
> thing. Will they not then believe? (Qur'an: 21:30)[32]

The verse points out that the universe was in a condensed state. In modern scientific terminology, that state is called singularity. The term refers to the moment when time and space merged into one entity. Astronomers use the term Big Bang to describe that initial defining moment of creation of the universe. Following that, God created life from water. "Every living thing (minal-maa-'i kulla shay-'in hayy)" in the above verse certainly includes mankind. Therefore, the early stage of the evolution of man occurred in water.

A sampling of some Qur'anic verse confirms the idea that humans are linked to both organic and inorganic realms:

> We fashioned man from quintessence (sulalah) of clay. (Qur'an: 23:12) [33]

> We fashioned you out of soil (turab). (Qur'an: 22:5) [34]

The word sulalah, in the Arabic text of verse 23:12, means "an extract of a thing; the clear, or pure, part or choice, or best, or most excellent part of a thing." [35] One may conclude that man was created from an extract of clay. Does sulalah mean silicone? We have learned previously that at least some scientists induct clay minerals as a catalyst, as well as a stabilizer, in the polymerization reaction of amino acids to RNA or DNA. The word Turab, which appears in verse 22:5, means "dust or soil." [36] Unlike clay, soil contains organic matter. Therefore, the Qur'an is telling us that organic matter or other life forms were created before man, and they were elements in the evolution of human beings.

Muslims believe that Qur'anic verses cannot contradict each other on any particular issue. The Qur'an states in various verses that man was created from water, clay, quintessence of clay, soil, and so on, and we know that these substances are different in chemical and physical constitution. Unless we consider these substances part of different stages in the early creation of humans, then the Qur'anic verses do contradict each other. If the prevalent contemporary Muslim belief in an instant creation with no intermediate stages is taken as the truth, the contemporary *Ullamah* (Muslim religious scholars) must propose credible explanations for these conflicting verses.

In their golden history and centuries before Darwin, Muslims believed that in the process of creation, man passed through a stage in the plant kingdom long before he appeared as Homo sapiens. The foundation for such a belief is: "And God has caused you to grow as a plant *(nabata)* from the earth." (Qur'an: 71:17) The Arabic verb *nabata*, used in the Arabic text of the verse, has a literal meaning, "to grow or germinate a plant," and a derived meaning, "to nourish a child."[37] Here, the literal meaning applies. The Kingdom Protista includes microorganisms that cannot be classified into animal or plant kingdoms. Could the verse be hinting at the Kingdom Protista as the classification through which all plant and animal kingdoms evolved?

Qur'anic verse 24:45 suggests that mankind is a part of the animal kingdom:

> And God has created *(khalaqa)* every living animal from water: Of them some that creep on their bellies: Some walk on two legs: Some walk on four: God created *(khalaqa)* what He wills: For verily God has power over all things . . . *(Wallaahu khalaqa kulla daaabbatin-mim-maaa': fa-minhum-mâny-yamshii 'alaa batnih; wa minhum-mâny-yamshii 'alaa rijlayn; wa min-hum-mâny-yamshii 'alaaa 'Araba'* . . .)[38]

Verse 21:30 uses the phrase "every living thing *(minal-maa-'i kulla shay-'in hayy)*." Here, verse 24:45 offers additional clarification: "every animal *(kulla daaabatim-mim-maaa)*." This verse specifies that animals were created out of water and explains the categories included among animals. Humans and sometimes apes are the only living creatures that always walk on two legs, while birds can fly and walk on two legs. Therefore the two-legged animals mentioned in verse 24:45 could be either bird or ape or human being.

The grammatical structure of the above verse with the phrase *fa min hum* is highly significant. If the noun *dabbah* (animals) were applied only to rational or irrational creatures separately, the two phrases, *fa min-hunna* or *fa-min-ha*, would have been used in proper Arabic grammar. Instead, the Qur'anic use of the phrase *fa min hum* in the verse conveys the Arabic noun *dabbah* (animals) in the verse refer to both rational and irrational creatures. [39] Therefore, the verse states that a rational animal that walks on two legs was also created from water. Human is the only creature that walks on two legs all the time. Moreover, human is most rational of all creatures. Therefore, the rational animal that walks on two legs referred in the verse is human and so, human belongs to animal kingdom.

Moreover, five hundred years before Darwin, the Muslim scholar, Ibn Khaldun wrote, "[M]an belongs to the genus of animals" and that "God distinguished from them by ability to think, which He gave man and through which man is able to arrange his actions in an orderly manner."[40]

The Qur'an addresses mankind, proclaiming that God created both mankind and also "those who came before" mankind:

> Oh! Mankind! Worship your Lord, Who hath created
> you and those before you, so that ye may ward off (evil).
> (Qur'an: 2: 21)[41]

Hence, it tells us that there existed other people before the emergence of modern man (Homo sapiens). We call them Homo erectus, Homo ergaster, Neanderthal man, Cro-Magnon man, etc., but, six hundred years before Darwin, the great Muslim Sufi, Ibn Arabi, called them "the animal men." Most Muslims interpret the above verse in a more limited fashion, believing that the term "mankind" means people living at the time of the Prophet, and that "the people before you" refers to the ancestors of the contemporaries of the Prophet.

In the next chapter, Qur'anic evidence is presented to document that Adam and Eve were the progeny of a different human species. According to the Qur'an, man is only a part of the grand design of God. Furthermore, according to the following verses, God may create new species transfiguring Homo sapiens:

> You shall surely travel from stage to stage. (Qur'an: 84:19)[42]

> We may transfigure you and make you what you know not. (Qur'an: 56:61)[43]

Additionally, the following verse predicts a newly created species able to verbally communicate with human beings:

> And when the word is fulfilled concerning them (mankind), We shall bring forth a creature of earth to speak unto them because mankind has no faith in Our Revelation. (Qur'an: 27:82)[44]

Clearly, the following verse indicates that the survival or extinction of any life forms is subject to God's will:

> Thy Lord does create and choose as He pleases. (Qur'an: 28:68)[45]

Geology and paleontology occupy a key position in evolutionary studies. We have quoted earlier, in 29:20, that the Qur'an counsels Muslims to study the earth and its contents to learn how God's process of creation originated and developed. The verse encourages mankind to learn to create, and we have earlier offered historical documentation showing that Muslim scholars undertook the study of geology and fossils long before Europeans started to do so. We have seen that al-Biruni observed that some hominid species were much bigger in size than modern man. He based his conclusion on his observation of the size and construction of ancient caves and the bones buried in them.

One can find many points of harmony between modern science, the Qur'an, and these early Muslims: First, the Qur'an and at least some scientists agree that clay played the role of a catalyst in the origin of life and man, and that life began in water. Second, they agree that life is a bush with many branches, and man is only a small twig on that bush. Third, they both confirm that the origin of life and its development has been a gradual process, not an instantaneous event in our time frame.

Both geology and paleontology provide information on the origin of life and its progressive changes, and they declare that human beings belong to the animal kingdom, that the original life form was gradually transformed to new forms, and that this transformation occurs with the participation of "the office of active nature (kiyan)." They also suggest that, in the future, human beings may be transfigured into a different form of life, the shape of which is not known to us at this time.

The following verses connect all the above verses together to show how God bound all life forms together into a whole:

> Extol the limitless glory of thy *Raab*: [the glory of] the All-Highest, who creates (everything), and thereupon forms it in accordance with what it is meant to be, and who determines the nature (of all that exists), and thereupon guides it (towards its fulfillment). (Qur'an: 87:1-3)[46]

In the above verses, the Qur'an commands the Muslims to glorify their *Rabb*, who creates everything and then lets His creations advance to perfection under His guidance. The word guidance has to be understood within the context of Islamic metaphysics of the future. Within this metaphysics, guidance means that God provides possibilities or proposals to each stage or component of the universe from which to choose and by which to realize their next state. All changes at every level are gradual in our framework of time and space. Creatures' selective actualization of the options suggested by God results in visible monuments of divine creation. (See CHAPTER 1 on the Islamic metaphysics of the future.) When genes accept a suggested possibility or possibilities from Allah, a new species is born or created. We will discuss this aspect in detail later.

Within the logic of this metaphysics, we can assume that variations of natural species that appear seemingly without design are in fact the results of divine will and grand design. This is not an attempt to fit the Qur'an to modern scientific reasoning; on the contrary, nine hundred years before Charles Darwin, al-Biruni understood the ways of God and how He brings about changes in life forms, as presented above. In Athar-ul-Bakiya (Chronology of Ancient Nations), he asserted that natural changes in life forms are decreed by Divine will and "show that the Creator, who had designed something deviating from the general tenor of things, is infinitely sublime, beyond everything which we poor sinners may conceive and predicate of Him."[47]

Furthermore, al-Jahiz (776-869) wrote:

> Individual defects in Nature do not lead us to the conclusion that all natural things happen by accident and chance. Thus the claim, founded on individual cases, that things happen by accident is incorrect and betrays ignorance. If you object, 'Why should such cases occur?' we reply that things are not determined by Nature so that the only explanation is, contrary to the claims of some people, that they are permitted by design of the Creator. He made the Nature run its normal course in most cases, deviating sometimes, however, because of certain accidents; this fact implies that Nature is commissioned and designed, needing the will of the Creator to reach its purpose and achieve its works."[48]

In modern scientific terms, the description of "something deviating from the general tenor of things" by al-Biruni or "Nature . . . deviating sometimes" by al-Jahiz can be read as the process of mutation. Later we shall discuss how the elements of chance and accidents merge with the creation.

All life forms on earth, including virus and vampire, man and monkey, grass and grasshopper, are the end result of the difference in the number and sequence of the four basic nucleotides (adenine, cytosine, guanine, or thymine) made out of one nitrogenous base, one 5-carbon sugar (deoxyribose), and one phosphoric acid residue. The genetic structure of the chromosomes determines the minutest physical, chemical, and immunological characteristics of various species in both the animal and plant kingdoms. Adding the meanings of *Rabb, al-Khaliq, al-Baari,* and *al-Musawwir* to our knowledge of paleontology, homology, analogy, molecular biology, and other branches of science, we may conclude the following about the matter of creation: God, through His proposals, proportioned and shaped the number and sequence of the four nucleotides in the genes to create new forms of life from previously existing creatures.

The Qur'an commands Muslims to learn the about process of creation by observing and decoding its visible signs on the earth and in their own selves (Qur'an 51:20-21 and 29:20). The early Muslims followed these commands in their exploration of nature. In classical times, centuries before Darwin, Muslims believed that the creation of mankind began with clay and water, and progressed through the lower life forms to Homo sapiens. Western scientists have only recently accumulated additional empirical data from the earth (fossils), from morphology, from the biochemistry of all organisms, and from other sources. This data supports Muslim thought on how God makes (*khalaqa*) new life forms from nonexistence. Therefore, observation of the evolution of life and the Qur'anic teaching of creation by God are not contradictory but complimentary. Unfortunately, Western scientists who grew up in Judeo-Christian traditions and who did not agree with the literal meaning of Genesis were forced to interpret the scientific data differently.

The study of comparative anatomy led to the concept that there are two types of similarities between living organisms. The first is analogous resemblance, where a fundamentally dissimilar structure has been modified or adapted to serve similar ends. For example, the resemblance between the surfaces of the wings of a fly, a bird, or a bat is analogical, but they are formed from different materials. The insect wing is simply a membrane connected by veins, while all vertebrate wings are created from the typical bones of the forelimb of a land vertebrate. The second, homologous resemblance, is when a fundamentally similar structure is modified to serve dissimilar ends. The hind limbs of a horse, a bat, or a whale, have bones that are modified in size, in details of shape, by reduction, or even by fusion of bones (as in the horse) to serve the needs of the different species. (See Figure 5-3.)

Evolutionists point out that analogy and homology provide unequivocal evidence that evolution occurs without supernatural intervention. But if read in the light of the literal meanings of the Arabic verb *khalaqa* and God's epithets *al-Baari, Rabb,* and *al-Musawwir,* analogy and homology would point to a divine act of creation by proportioning one thing to another.

Creation and evolution are integral parts of a divine act. Interpretations of analogy and homology vary depending upon the interpreters. Atheistic evolutionists consider them evolutionary processes detached from divine intervention, while Muslims, in the classical period, considered them willful acts of divine creation. In the light of Islamic metaphysics of the future, we can see analogy and homology as the results of some creatures' acceptance of divinely suggested possibilities arriving through the messenger moments of the future.

One of the cardinal principles of the modern synthetic theory of evolution is the slow accumulation of small mutations. In excerpts quoted in the previous chapter, Ibn Khaldun expresses the cardinal principle of Neo-Darwinism in terminology used in his time. He describes evolutionary processes as the "connection and full preparation to become the next stage," "the last stage of each group is fully prepared to become the first stage of the next group." Such statements can be paraphrased using modern synonyms as "the accumulation of small mutations results in speciation."[49]

When Judeo-Christian theologians started to present evolution as a divine plan slowly unfolding through the ages, Bertrand Russell, a pacifist philosopher, wondered whether, during such process, "the omnipotence (God) [was] quietly waiting for the ultimate emergence of man?"[50] Similarly, many traditional Muslims question the truthfulness of the stage-by-stage progression of creation because they observe the literal meaning of the Qur'anic verse: "Verily, when He intends a thing His command is 'Be', and it is." (Qur'an: 36:82)[51]

The theory of relativity provides answers to these important concerns. Einstein convinced the world that time is not absolute, but relative. He demonstrated that each observer carries around his/her own personal time, and it does not generally agree with anybody else's.[52] Time is not absolute; it is elastic. (Refer to chapter 2). Therefore, for God, our experience of events and characters in the drama of life is compressed into "a twinkling of an eyelid." This is depicted in Figure 7-1.

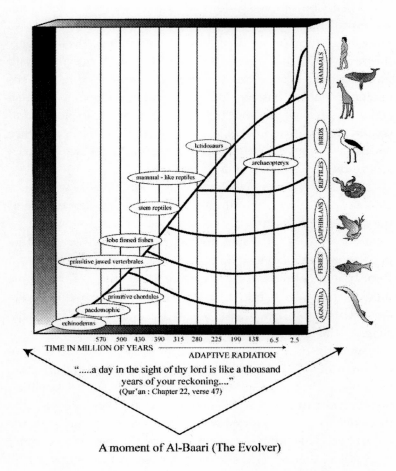

A moment of Al-Baari (The Evolver)

Figure 7-1. EVOLUTION, CREATION, THE THEORY OF RELATIVITY, AND AL-BAARI (THE EVOLVER).

In the ninth century, al-Biruni argued that the age of the universe could not be calculated. He asserts, like our modern physicists, that for an observer on earth, "the colossal and minute moments of creation seem to have formed over a long period of time."[53] The events in the process of the creation of life are current events in the time frame of God, but the earthbound man experiences creation as a multitude of events and evolution occurring over a period of 3.5 billion years.

When God orders Muslims to "travel around the earth and see how [He] originated creation" (Qur'an: 29:20), we do so knowing that the process of creation is comprehensible to us only through our earthly experiences. Had God said one thing and given evidence of another, we would have cause for doubting His existence, but He has provided us with evidence, tools, and an ability to discover and understand His creation. By exploring and comprehending that process, we confirm our belief in His existence and are able to accept the assertion that human existence occurred as a result of His command.

Sadly, many contemporary Muslims do not realize when they view life as a movie that they are in it. They may accept God as the Creator, but assume that their creation has no connection with any primates. They may not realize that their lives constitute one frame of the movie of life, and that there are other frames in front of and behind them. They claim that they are the whole movie. They begin and end with themselves. However, God's epithet, the Evolver (al-Baari), within space-time relativity, suggests the opposite. (Figure 7-2.)

Similarly, materialists among scientists, who can see most of the movie of life, are satisfied with the concept of natural selection to explain creation. We have seen earlier that Pre-Darwinian Muslims described an active nature (kiyan). They believed that the active nature is endowed with a faculty to select species as the farmer selects the better seed. They believed also that its laws submit to the will of God. Materialists accept the existence of natural laws— but they evade the question of why natural laws exist.

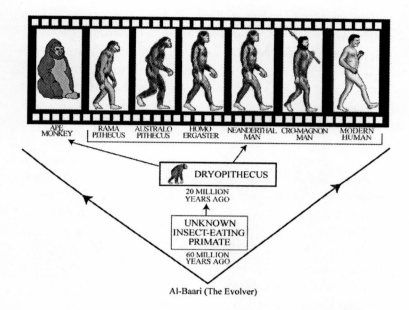

Al-Baari (The Evolver)

Figure 7-2. *AL-BAARI* (THE EVOLVER), SPACE-TIME RELATIVITY, AND THE EVOLUTION OF MAN FROM AN INSECT-EATING MAMMAL.

When we look around us, we know where tangible things are—where the United States is, where Mecca is. We recognize the position of moving objects (cars, trains) at any particular moment. When we first extended our experience into the atomic world and drew a picture of an atom as a micro-version of our solar system, we assumed the internal constituents of the atom must rotate just as precisely as the planets around the Sun. However, we now see that when we descended the stairway into the atomic world, we lose our sense of certainty about everything and find ourselves in a fuzzy and chaotic world.

An electron does not appear to follow a meaningful, well-defined trajectory at all. All known subatomic particles—even whole atoms—"cannot be pinned down to a specific motion. Scrutinized in detail, the concrete matter of daily experience dissolves in a maelstrom of fleeting, ghostly images."[54] This uncertainty is the fundamental ingredient of quantum theory. It leads directly to the consequence of unpredictability. "Most scientists . . . accept that the uncertainty is truly intrinsic to nature [and that] when it comes to atoms, the rules are those of roulette."[55]

Electrons and other particles simply pop about at random, seemingly without rhyme or reason. When we look for the location of an atom, we see an atom at a place. When we look for its motion, we get an atom with a speed. But we cannot fix on both factors at the same time. Thus, the world of atoms and subatomic particles is an uncertain and unpredictable world. Biologists who reject the divine connection to the origin of the species are like atoms living in their uncertain, chaotic universe. Atomic particles are incapable of imagining, based on their limited experience, that a predictable nature and a coherent universe can exist above all the uncertainties of their "colossal" micro-world.

Divergence between materialist biologists and the Qur'an is not over proven facts but over inferences drawn from these facts. Biologists infer that gradual changes in life forms are due to the cumulative mutations dictated by environmental changes alone. They do not believe in the need to invoke a supernatural power to explain life and its origin. On the other hand, the Qur'an declares that God created all forms of life, with their deficiencies and adaptabilities, in the environments in which they exist. It maintains that the extinction of old species and the appearance of new ones are governed by divine laws and by the arrival of future moments packed with optional possibilities for God's creatures with free will to actualize their selected options into monuments of divine creations. Muslim scholars, long before Darwin, believed that environmental factors cause changes in life forms, but they also held that such factors are subject to God's guidance. For example, Ibn-Khaldun, after attributing the origin of human races to environmental factors, corroborated his point by quoting the Qur'anic verse: "This is how God proceeds with His servants. And verily, you will not be able to change God's ways."[56]

Another discrepancy between scientists and the Qur'anic teachings is that the latter asserts that the highest and most perfect form of God's creation occurred when the Divine Spirit *(Ruh)* breathed into Adam and Eve's animal bodies. The Qur'an explains the word *Ruh* in this way:

> And they ask you about soul *[Ruh]*. Say: The soul is one of the commands of my Lord, and you are not given aught of knowledge but little. (Qur'an: 17:85)[57]

The Qur'an states:

> And when your Lord said to the Angels: Surely I am going to create and shape a mortal from the essence of black mud. When I have made him complete and breathed into him My Spirit *[Ruh]*, fall down and make obeisance to him. (Qur'an: 15:28-29)[58]

Thus, while materialists and the Qur'an share the concept of the evolutionary progression of earthly existence, they make different inferences as to the cause of such progression.

We have surveyed statements in the Qur'an that point us to the origin of humanity. Scientists and Muslims of our time might say that I am trying to reconcile the Qur'an with modern science, but data presented earlier confirms that many early Muslims believed in the origin of life as described in this chapter. Centuries before Darwin, Ibn Miskawayah, Ibn Khaldun, Ibn Arabi, al-Biruni, Ibn Sina, and other medieval Muslims taught that man has an ancestral link to the ape; yet such information was shocking news when Darwin conveyed it to the Judeo-Christian West. I do not need to fit the Qur'an into modern scientific beliefs because, conversely and unknowingly, modern science confirms the accuracy of the Qur'anic verses and what early Muslims wrote centuries ago.

Finally, the Qur'an reveals, and the pre-Darwinian Muslim scientists concur with, the concept of "creation by modification" rather than "the descent by modification." This modification occurs through mutations in the genes, which, in our time frame, may have taken place over a period of four billion years, but in God's time frame, happens in no time and is a current event.

In summary, Muslims believe that the natural forces that impacted life and vice versa are caused by divine laws, while modern scientists believe that these natural laws are the original and final answer to all of our questions, despite the fact that they do not explain how natural laws came into existence. Additionally, Muslims believe that, although natural laws and events are comprehensible, God is incomprehensible. The Qur'an reads:

> No vision can grasp Him, but His grasp is over all vision:
> He is above all comprehension, yet is acquainted with
> all things. (Qur'an: 6:103)[59]

In our survey of early Muslim thought and the study of the Qur'an, we have learned that Muslim scholars described the evolutionary process in detail. Although they and the Qur'an did not use the terminology used by scientists today, they believed in the ever-growing bush called life long before modern scientists such as Stephen Jay Gould or Ernst Mayr.

Not only did Western scholars hijack the fundamentals of the Muslim theory of evolution and present it as their own pristine idea, but, knowingly or unknowingly, they also excluded Muslims from their archives. For example, *The Encyclopedia of Evolution* by Richard Milner, foreword by Gould, excludes Muslim authors from the history of the theory of evolution.[60] That omission is comparable to writing the history of the United States without mentioning George Washington, John Adams, Thomas Jefferson, and Abraham Lincoln.

CHAPTER 8

The Creation of Adam and Eve from the Animal World

In the last chapter we learned from the Qur'an that the creation process is evolutionary and that many medieval Muslim scholars held the view that humankind evolved from the animal world. However, most *contemporary* Muslims believe that Adam was created *ex nihilo* and that Eve was created from his rib. In this chapter we will explore the Qur'an to choose between the two beliefs.

> O you man, what deluded you concerning your Munificent Lord, who created you {خلق (qhalaqa)} fashioned you {سوّ (sawwa)} and shaped you in perfection {عدل (hadala)}. (Qur'an 82:6-7)[1]

> And We who created {خلق (khalaqa)} you, and fashioned {سوّر (sawwara)} you, then told the Angels: fall ye prostrate before Adam! And they fell prostrate, all save Iblis, who was not of those who make prostration. (Qur'an 7:11)[2]

If God created Adam with no connection to the animal world and other hominids, the Qur'an would not have said, "We created you, then fashioned you." Similarly, if Adam did not evolve in stages but was a perfect creation, then the verb to sculpt (*sawwa or sawwara*), which appears in the verses, becomes superfluous because there is nothing to perfect in a perfect being. The commentator translates the verb *hadala* (عدل) as "shaped you in perfection." Lane's lexicon gives another meaning to this word: "rate a thing as equal to a thing of another kind so as to make it like the latter." Therefore, it means: transform a thing into another, so that latter is distinctly identifiable from the former. In support of this meaning, he quotes the following Qur'anic verse: "All praise be to God who created the heavens and the earth, and ordained darkness and light. Yet the unbelievers make others equal *{hadala* (عدل)*}* of their Lord (Qur'an 6:1)". Lane gives the phrase عدّل بفلان فلان (*phulaanan biphulaanin hadala*), which means, "He made such a one to be equal, or like, such a one."[3] Therefore, the verses 82:6-7 must be rephrased as: "O you man, what deluded you concerning your Munificent Lord, who shaped you from a pre-existing thing (*khalaqa*), sculpted you (*sawwa*), and transformed you to [distinctly identifiable] perfect human shape [i.e., Homo sapien sapiens] (عدل *hadala*)." Similarly if we replace the words "created and fashioned" with the meaning of *khalaqa* and *sawwara* of the classical Arabic, verse 7:11 would read: "And We who shaped you from a pre-existing thing (*khalaqa*), and sculpted or perfected you to become distinguished with modern human (first Homo sapien sapiens or Adam and Eve) characteristics (*sawwarra*), then told the Angels: fall ye prostrate before Adam! And they fell prostrate, all save Iblis, who was not of those who make prostration."

We have learned earlier the divine epithet *al-Musawwir* (sculptor) could also mean, in current scientific vernacular, someone who creates a distinct species.[4] Therefore, *al-Musawwir* created Adam and Eve, who were the first perfect human beings (Homo sapien sapiens) sculpted from a previously created hominid species. This early hominid was identified by Ibn-Arabi as an animal-like man, but modern biologists called him the Neanderthal or Cro-Magnon.

Adam and Eve could have been born as progeny of this early

species, and one can arrive at such interpretation from the following verses:

> Behold, thy Lord said to the angels: 'I will establish a successor (*khalifah*) on earth.' They said: 'Will Thou place on it such as will spread corruption thereon and shed blood—while we celebrate thy praises and glorify thy (name)?' He said: 'I know what you know not.' (Qur'an 2:30)

Most Muslim commentators of the Qur'an translate the noun *khalifah*, in the above verse, as vicegerent, but it is actually derived from the verb *khalafah*, which means "one entity succeeds another or remains after another that has perished or died."[5&6] Therefore, the literal translation of the noun *khalifah* is successor, but the term vicegerent is also correct, because human supremacy over the earth was established by divine revelation of all knowledge to Adam and Eve. They are the first perfect Homo sapiens that "succeeded or remained after" the earlier inhabitants of the earth "that had perished or died." We have paleontological data that pre-modern human species existed and perished with the emergence of modern humankind. Therefore, the use of the word *khalifah* in the above verse suggests that Adam and Eve were transformed (*khalaqa*) or evolved from earlier species.

Verse 6:133 gives explicit evidence that modern man (*Homo sapien sapiens*) was evolved from earlier hominids.

> Thy Lord is All-sufficient, Merciful. If He will, He can put you away, and leave after you, to succeed you, what He will, as He produced you (انشأ) you from the seed of another people. (Qur'an 6:133)[7]

The context in verse 130 of the same chapter shows that it is addressed to mankind. Almost all translators of the Qur'an render the Arabic word '*ansha* (انشأ) as *raised*. This is not a correct translation of the word, however. As per the above translation,

Edward Lane's *English Arabic Lexicon* translates it as follows: "originated it; brought it into being or existence; made it, or produced it, for the first time, it not having been before."[8] Initially, verse 6:133 states that God may replace modern human with another species ("whomever He pleases to succeed you"). The process of creation of the new species (*khalifah*) from human seed is similar to when God "brought humanity into existence for the first time (انشأ)." The verse also suggests that mankind was originally created from "the seeds of another people," i.e., that Adam and Eve were evolved from earlier people. Therefore, verse 6:133 can be rephrased as: "Thy Lord is all Self-sufficient, Merciful. If He will, He can put you away, and leave after you, to succeed you, what He will, as He originally created (انشأ) you [as species] from the seed of another people."

Who are these other people? As already discussed, Ibn Arabi labeled them "animal man." Modern biology identifies them as Homo ergaster, Neanderthal, Cro-Magnon, etc.

> God did choose Adam and Noah, the family of Abraham, and the family of Imran for high distinction. The one is the progeny of the other God heareth, knoweth. (Qur'an 3:33-34)[9]

In the light of verse 6:133, the phrase "offspring, one of the other (*zurriy-yatam-ba'-zuhaa mim-ba'-z*)" in verse 3:34 is binding on Adam, as the context clearly shows, and supports the theory that Adam and Eve were the offspring (*zurriyah*) of previously created species. Adam and Eve had no Homo sapien sapiens parents but, genetically speaking, they were mutant variants or "offspring" of previously created species from which they were created and fashioned (shaped). This assertion is in complete agreement with the teachings of Muslim scientists of the classical period of Islam.

A contemporary of Charles Darwin, Professor John William Draper (1812-1883) of New York University, admitted that Muslims in their classical times believed in divinely ordained transmutations of species. He writes: Transmutation and transubstantiation were two sisters, destined for a worldwide celebrity; one became allied with the science of Mecca, the other with the theology of Rome.[10]

As was mentioned earlier, most Muslims of our time believe that Eve was created from Adam's rib. They point out that the following Qur'anic verse supports such belief:

> O Mankind! Be conscious of your Sustainer, who has created you out of one living entity (min-nafsin), and out of it created its mate, and out of the two spread abroad a multitude of men and women. (Qur'an 4:1)[11]

As the verse states that the mate was created from a previously created single entity (nafs), many Muslims believe that the first soul was Adam and the mate was Eve. Abul A'la Maududi, a well-respected Islamic scholar of our time, comments on this verse: "At first one human being was created and then from him the human race spread over the earth." Maududi observes that the Qur'an reveals to us that the "single soul" was Adam and that mankind originated from him, but the Qur'an does not give us specific knowledge as to how Adam's mate emerged from him. He points out that the story that Eve was created from the rib of Adam is adhered to not only by Muslim commentators, but has also been revealed in the Bible and the Talmud. Maududi then points out: "the Qur'an is silent about it and the Tradition of the Holy Prophet that is cited in support of [it] has a different meaning from what has been understood." He advises us "to leave it undefined as it is in the Qur'an and not to waste time in determining its details."[12] If Muslims follow Maududi's advice, they would be disobeying one of the Qur'anic commandment, which reads: "Travel through the earth and see how God did originate creation." (Qur'an 29:20). The verse admonishes Muslims to discover the details of the creation of man and the universe.

The word nafs in the above verse may mean soul, spirit, mind, animate being, living entity, human being, person, self, life's essence, vital principle, etc.[13&14] Most Muslim commentators have assumed that the original individual, *nafs,* is Adam and the mate (*zawj*) is Eve. Muhammad Abduh (1849-1905), a prominent Islamic scholar of Egypt, accepted "human being" as the meaning of the word *nafs,* but he did not agree with interpretations that Eve was created from Adam's rib. Muhammed Asad defines *nafs* as a "living entity." His choice is rational and also harmoniously meshes with the theories of classical Muslim scientists, who believed that humankind was created from the world of apes.

If modern humans were created out of other species (like the Animal Man), the parent embryo had to be transmuted into a Homo sapien sapiens embryo (modern human *nafs*). Therefore, it is logical to hypothesize that the first human being evolved through the transmutation of a pre-human embryo into a modern human embryo (modern human *nafs*) and from that the first Homo sapien sapiens discordant twins (Adam and Eve) were born. The Qur'anic verse "we created you out of one living entity *[min-nafsin],* and out of it created its mate" can be understood in that context. In other words, Adam and Eve were the first discordant human twins born out of Ibn-Arabi's Animal Man or modern scientists' Neanderthal.

In nature there are two kinds of twins—dizygotic and monozygotic. Dizygotic twins develop when two different sperm cells fertilize two ova (eggs). In monozygotic twinning, a single sperm cell fertilizes a single egg. Monozygotic twins emerge from the separation of a single developing embryo into two independent parts. Monozygotic twins are almost always of the same sex. However, the Qur'an states in verse one of chapter four that the first human being was created from a single living entity (*nafsin wahidatin*) from which God created a mate (*zawj*). Therefore, Adam and Eve were created from a "single living entity (*nafssin wahidatin*)." The Qur'an also states that humans originated from a single male and female. Therefore, Adam and Eve were monozygotic twins, and a transmutation occurred in the sex chromosome of one of the twins, making it into a "mate (*zawj*)" for the other. This unusual and deviant natural event (some people may call it a miracle) occurred due to an intentional command from God, as it happened in the case of the creation of Prophet Jesus.

One of the major foundations of the theory of evolution is beneficial mutation. No new species will ever evolve without mutation within the parent chromosomes. If mutation in the parent stock can produce new species, a mutation in one of the monozygotic embryos (zygotes) can also lead to monozygotic discordant twin births.

In 1961 the authors of *Presomption de Monozygotisme en Dipit d'un Dimorphisme Sexuel: Sujet Masculin XY et Sujet Haplo X,* reported such a pair of monozygotic twins of opposite sex.[15] The idea of sex change in the embryo by induced mutation is testable. Our contemporary science of molecular biology confidently states that we can select the sexes of babies to be born. Therefore, my theory, based on the Qur'an and the findings of early Muslim scientists, is not a tautology. I speculate that, in the future, this will be proven when an implanted in-vitro-fertilized and sex-selected female embryo results in discordant twins.

After showing that the origin of the races of mankind is a result of natural processes, Ibn-Khaldun states: "This is how God proceeds with His servants. And verily, you will not be able to change God's ways."[16] The origin of the different human races was the result of the mutation of genes within human chromosomes. For those who do not believe in God, it is merely a natural process, but for Ibn Khaldun and other Muslims of the classical times, it is a comprehensive natural phenomenon that is divinely ordained. Similarly, we can say that Adam and Eve were the products of a process of transmutation and twin births.

A significant reason for assuming that Adam and Eve were twins as described above is that the viable mutation with major morphological and physiological effects is extremely rare. The possibility of two identical rare mutant individuals with opposite sexual characteristics originating at the same place and at the same time is slim.[17] So, for the preservation of a new species (in this instance, modern humankind), a male and a female would have to be born simultaneously at the same place. Statistically it is more likely for a mutant discordant twin birth to survive as a species than it is for a single mutant Homo sapien sapiens.

In summary, Adam was not created *ex nihilo,* but from an earlier species that may take the modern name *"Homo ergaster"* or the name "Animal-type Man" of Ibn Arabi. This being, we believe, is the forerunner of Eve and Adam. The story that Eve was created from Adam's rib was a later addition to Muslim beliefs. The Qur'anic verse 4:1 suggests a twin birth for Adam and Eve.

Islam's greatest historiographer, Abd-ar-Rahman Muhammed ibn-Khaldun, believed that the creation of mankind was gradual and that it evolved from the world of apes. In an earlier section of this work, we have seen many other pre-Darwin Muslims echoing the same belief. How, then, did the Judeo-Christian belief that God created Eve from Adam's rib become a part of the Muslim faith? The *Muqaddimah* supports the argument that the story was not based on the Qur'an, but was incorporated into Muslim belief through Jewish and Christian converts to Islam.

Ibn-Khaldun maintains that the sources for this story can be traced back to early Muslims, contemporaries of the Prophet Muhammad and to men who belonged to the generation that succeeded him. Records preserved by these men involved both reliable and unreliable materials. The reason for that, according to Ibn-Khaldun, is that the early Arabs "had no books or scholarship [and] desert attitude and illiteracy prevailed among them. When they wanted to learn certain things that human beings are usually curious to know, such as the reasons for existing things, the beginning of creation, and secrets of existence, they consulted earlier People of the Book (Jews and Christians)." Ibn-Khaldun maintains that when these men converted to Islam, they clung to some of their Judeo-Christian beliefs, such as the beginning of creation. He asserts that converts such as Ka'b ul-ahbar, Wahb ibn Munabbiah, Abdullah Ibn Slam, and others "filled the Qur'anic commentaries with such materials, which originated . . . with the people of the Torah," and therefore such information was neither sound nor verifiable. In Ibn-Khaldun's view, their interpretations were accepted because they were "people of rank in [their] religion and religious community."[18]

Ibn-Khaldun states that God created man in a gradual manner, developing from minerals into plants and then into animals in a hierarchical manner by which the superior species always presupposes and rises above that which is inferior. Consequently his investigations led him to conclude that modern humans were evolved through transmutation from monkey and Animal Man. He further explains such transformation as follows: "The essences at the end of each particular stage of the worlds are by nature prepared to be transformed into the essences adjacent to them, either above or below."[19]

In the light of the testimony of Ibn-Khaldun, one of Islam's greatest historiographers and a devout Muslim, we can see that Muslims incorporated the story of the creation of Eve from Adam's rib not through the authority of the Qur'an, but through information they acquired from Jewish and Christian converts. There is no Qur'anic, historical, or scientific basis to accept the contemporary Muslim belief that Adam and Eve were created *ex nihilo*.

The Qur'an tells us that all races, tribes, and nations are the children of Adam and Eve. God addresses us as such: "O mankind! We created you from a single (pair) a male and a female, and made you into nations and tribes, that you may know each other. (Qur'an: 49:13) [20] He also tells us: "Mankind was one single nation." (Qur'an: 2:213)[21] Modern scientific research has come to the same conclusion.

Research by geneticists Rebecca Cann of the University of Hawaii, Mark Stoneking of Pennsylvania State University, and the late Allen Wilson of the University of California suggests that modern human beings evolved in Africa. These geneticists studied mitochondria in order to trace the origin of humankind. Mitochondria are thin bundles of enzymes floating in our cells that convert chemicals to energy. These enzymes contain very small amounts of DNA (deoxyribonucleic acid). An ovum (female egg) carries the mitochondrial DNA, not a sperm cell. Therefore, our mitochondrial DNA comes only from our mother. The woman's mitochondrial DNA passes intact to her daughters, granddaughters, and so on, except for a constant rate of mutation in it over a period of time. From this simple mode of inheritance and a constant rate of mutation, geneticists derived a "tree of descent that showed more differentiation in Africa than anywhere else. That finding indicated that the human mitochondrial DNA has been evolving for the longest time in Africa—that is, it can be traced to a single African woman."[22]

This "single African woman" could be the pre-Homo sapien sapiens woman from whom Adam and Eve were created, or could be Eve herself. But the scientists, who are so afraid and reluctant to associate science with any theological thought, in a second breath, state: "In fact, we have no evidence that there ever was a time when only a single woman lived on earth. Many other women might have lived at the same time, but their mitochondrial lineages simply went extinct."[23]

Another group of evolutionists, who reject the "Out of Africa Model of Human Origin and Eve Hypothesis," cites mathematical flaws in the construction of the hypothesis. In contrast to the "Out of Africa" model of human origin, the "Multi-regional" model says that modern humans evolved simultaneously from Homo erectus in many regions of the world. University of Michigan anthropologist Milford Wolpoff, the foremost champion of the multi-regional model, states: "If the mitochondrial clock has any validity at all, it is simply measuring the original migration of Homo erectus out of Africa. The Eve hypothesis is refuted. Case closed."[24]

However, the "case" has not closed yet. Recently, genetic researchers at Stanford University, the University of Arizona, and the University of Paris used the variations in the Y chromosome (which only males have) to trace the human lineage to the father of us all (Adam). Most chromosomes exist in pairs, so that during the scrambling of the chromosomes in the formation of a sperm cell, chromosome pairs can exchange chromosomal markers within it. The Y chromosome escapes this scrambling because it lacks a particular counterpart, so it is transmitted intact from father to son, over generations. Based on this premise, geneticists traced certain markers of the Y chromosome to its origin. An article in a leading journal in the scientific field, *Science*, summarised these studies as follows:

> "In the beginning, there was mitochondrial Eve—a woman who lived in Africa between 100,000 and 200,000 years ago and was the ancestor to all living humans. Geneticists traced her identity by analyzing DNA passed exclusively from mother to daughter in the mitochondria, energy-producing organelles in the cell. To test this view of human origins, scientists have been searching ever since for Eve's consort: Adam, the man whose Y chromosome (the male sex chromosome) was passed on to every living man and boy. Now, after almost a decade of study, two international teams have found the genetic trail of Adam, and it points to the same time and place where mitochondrial Eve lived."[25]

This scientific evidence does not contradict the Qur'anic assertion that humankind was created from a "single pair of a male and a female." While critics of the "Out of Africa" model point to mathematical flaws in the computer program and press for more evidence, Blair Hedges, evolutionary biologist at Pennsylvania State University, finds that the current statistics are powerful enough to avoid such flaws.[26]

Another exploration into the roots of modern humankind tracked its linguistic origins. Investigation into the origins of at least five thousand languages spoken today led to the conclusion that the root of all languages was to be found in Africa. The migration of people from Africa led to the evolution of languages.

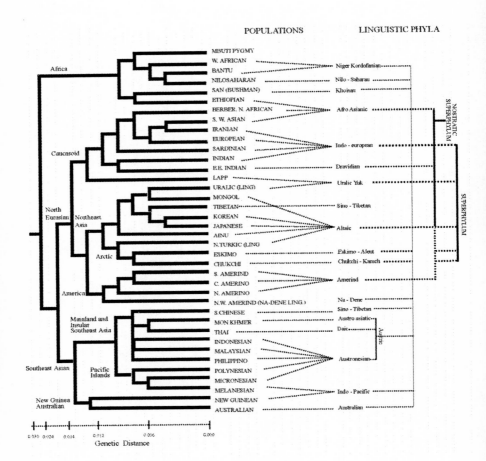

Figure 8-1. CORRELATION OF THE GENETIC TREE OF PEOPLE AND LANGUAGES. Linguistic super-families show an exceptionally close match with two major clusters, indicating parallel evolution of genetic and linguistic characteristics of the population. Source: Luigi Luca Cavalli-Sforza, Albert Piazza, Paolo Mennizi, and Joanna Mountain. Proc.Natl.Acad.Sci, USA. (Vol. 85, p. 6003. August 1988.) Reprinted with permission of Luigi Luca Cavalli-Sforza.

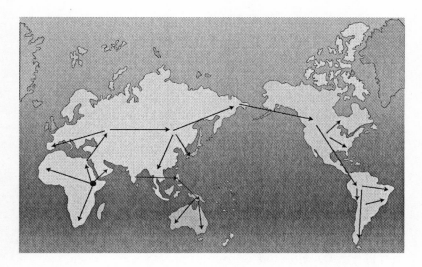

Figure 8-2. MIGRATION OF EVE'S PEOPLE FROM AFRICA. 200,000 years ago modern humans evolved in Africa, migrated to other parts of the world, and took control of the earth, exterminating all other previously existing human species.

"Human evolution is punctuated by the splitting of populations into parts, some of which settle elsewhere. Each fragment evolves linguistic and genetic patterns that bear marks of shared branching points. Hence some correlation is inevitable (between genetic and linguistic evolution)."[27] The distribution of genes correlates remarkably well with that of languages[28] (Figure 8-1). Genes and fossils narrate corresponding legends. In Figure 8-2, the present scientific speculation on the route of migration of the people from Africa is depicted based on genetic and linguistic studies.[29]

Why should genetic and linguistic evolution correspond so closely? In Wolpoff's "Multi-regional" model, the genetic and linguistic correlation cannot exist because genes do not control languages. Therefore, genetic and linguistic correlation suggests the "Out of Africa" model.

We can argue that deciphering the linguistic and genetic correlation by wise women and men may be what the Qur'an alludes to when it says, "And among His Signs is the creation of the heavens and earth, and variations in your languages and in your colors: verily in that are signs for those who know." (Qur'an 30:22) [30] The Qur'an also tells us that mankind was once a single nation, and differentiation in color and language occurred later. The black, the white, and the brown are branches of the same tree whose roots are Adam and Eve. The literal meaning of the word "Adam" is a "person with a dark complexion who originated from Earth."

Based on these evidences and others, there are no contradictions between the "Out of Africa" model, the proclamation of the Qur'an, and the Pre-Darwin Muslim concept of creation.

Currently, the theory of evolution has become a kind of religion for the scientific community, but the origin of mankind from a single pair, one male and one female, is yet to be established fully. It is plausible that in the future, it may become part of the scientific theory on the origin of humankind. The theory of evolution developed by Muslims, centuries before Darwin, has become the predominant paradigm for today's modern biologists.

Many scientists reject the "Out of Africa" model of human origin and the discovery of mitochondrial Eve and Adam. However, it is unscientific to reject the theory of the African origin of humankind when different researchers come to the same conclusion using the same or different tools. When Darwin brought the theory of evolution to the attention of the West, they initially rejected it. We may have to wait a while to see the incorporation of Adam and Eve into a new paradigm of agnostic science.

CHAPTER 9

The Location of the Garden of the Forbidden Tree

Most contemporary Muslims across the world believe that Adam and Eve were created in Paradise (*Jennat-ul-Khuld*) but were expelled for eating fruit from the forbidden tree in the garden. Early Muslims carried on great debates about the location of the garden. According to the two foremost exegetes of the Qur'an, Ibn Kathir (died in 1372) and ar-Razi (died in 1209), four interpretations of the location of the garden prevailed: That the garden was Paradise itself; that it was a separate garden created especially for Adam and Eve; that it was located on Earth; and the view that it was best for Muslims not to be concerned with the location of the garden.[1 & 2]

Unorthodox as it seems for our time, more reasons lead us to believe that the garden was on Earth rather than in Paradise. Many Qur'anic verses suggest that Paradise is a place the eyes of believers have never seen:

> No living soul knows what comfort of eyes has been
> kept hidden from them as a recompense for their deeds.
> (Qur'an 32:17)[3]

Abul A'la Maududi, a well-respected twentieth-century commentator on the Qur'an, tells us that Bukhari, Muslim,

183

Tirmidhi, and Imam Ahmad, all most respected scholars in the field of the Prophet's tradition, cited Abu Hurairah's text (a contemporary of the Prophet) that the Holy Prophet said: "Allah says: I have prepared for my righteous servants that which has neither been seen by eyes, nor heard by ears, nor ever conceived by any man."[4] If no human eyes have seen it, then Adam could not have seen it, and therefore the Garden of the Forbidden Tree could not have been in Paradise.

The contextual analysis of following verses also rules out the possibility of the garden being Paradise:

> And He taught Adam the nature of all things; then He placed him before the Angels, and said: Tell me the nature of these if you are right . . . And behold! We said to the angels: bow down to Adam and they bowed down. No, said Iblis (Satan). He refused and was haughty. He was of those who rejected the Faith. (Qur'an 2:31 and 34)[5]

When Satan refused to prostrate before Adam, God ordered him out of the garden. God said:

> "Get thee down from this: it is not for thee to be arrogant here: get out, for thou art of the meanest (of creatures)." (Qur'an 7:13)[6]

Then in the presence of Adam and Eve, Satan requested and God replied:

> "Give me respite till the Day when they shall all be raised up from the dead." Allah replied, "You are granted respite." (Qur'an 7:14-15)[7]

According to the above verses, Adam learned the nature of all existing things (Qur'an 2:31). Therefore when God granted Satan respite, Adam and Eve must have known that their lives in the garden with the forbidden tree were subject to imminent death and that they would be "raised up from death" on the Day of Judgment (Qur'an 7:14-15). For these reasons, the garden could not have been in Paradise but must have been on earth. It would make no sense for Adam to be promised Paradise as a reward for his good deeds if he were already there. In fact, Satan deceived Adam and Eve: "Your Sustainer has forbidden you this tree, lest you two become [as] angels, or lest you live forever." (Qur'an 7:20)[8] He whispered to Adam: "O Adam! Shall I lead you to the tree of eternity and to a kingdom that never decays?" (Qur'an 20:120)[9] Thus, by describing the forbidden tree as a secret recipe for eternal life, Satan tempted Adam and Eve to partake of the fruit of the tree. When they consumed the forbidden fruit, they sought eternal life.

The Qur'an states that Paradise is eternal. If Adam and Eve were actually in it, then they too would be eternal. The desire to acquire a thing develops in the human mind only when the desired object or state is not available. Adam was in an environment where death and decay, not immortality, were the facts of life. Adam and Eve would not have become victims of temptation if they had already been immortal beings in Paradise.

The following verse adds more support to the falsity of the notion that the forbidden tree was in Paradise.

> And I have secured it (Heaven) against every rebel Satan.
> These satans cannot hear the words of the exalted ones;
> they are darted at and driven off from every side, for
> them there is a perpetual torment. However, if someone
> snatches away something, a flashing flame follows him.
> (Qur'an 37: 7-10)[10]

Abul A'la Maududi explains the above verse as follows: "The satans have no access to Heaven and no power to hear the angels' conversations . . . if by chance a little of its news enters the ear of a devil, and he tries to bring it down, he will be followed by a flashing flame."[11]

Based on verses 20:120, 7:20, 37:7-10, and 7:22, we can conclude the following: First, Adam was allowed to stay in the

Garden of the Forbidden Tree after Satan's expulsion from it. Second, although Satan did not have access to Paradise, he succeeded in entering and deceiving Adam. Third, Adam and Eve believed Satan and ate from the tree in the hope of becoming immortal. Thus, the Garden of the Forbidden Tree was a place where all that existed had a transient, decaying nature into which Satan was able to enter. How could it be the Garden of Eternity?

Qur'anic verses repeatedly disprove the idea that the garden was in Paradise.

One verse reads: "There [Paradise] they call for every kind of fruit in peace and security: Nor will they there taste death . . . ." (Qur'an 44:55-56)[12] Thus the presence of a forbidden tree bearing the fruit of immortality and the commandment that Adam and Eve not eat of that fruit contradict the location of the Garden of Forbidden Tree in Paradise.

Throughout the Qur'an, Paradise is described as a place where peace prevails: "There they shall hear no idle talk, but only 'Peace'" (Qur'an 19:62)[13], where sin, lies, and vanity do not exist (Qur'an 56:25-6; 78:35, and 88:11). [14] If we accept the Qur'anic description of Paradise, then we also see it could not be the place where Satan makes the false, vain claim that he is better than Adam:

> He (Iblis) said: "I am better than he [Adam]; You have
> created me of fire and created him [Adam] of clay."
> (Qur'an 7:12)[15]

If Satan's claim were true, God would not have asked the angels and the Jinn to prostrate before Adam—thus Satan lied. In Paradise, Satan would not have been able to tempt Adam to commit sin. Such egoism, lying, and deception could not have taken place in Paradise—a place where fruits are not forbidden, where peace and justice prevail, and where sin, lies, and vanity are unknown.

Finally, Adam could not have been thrown out of Paradise because it is the place of final return: " . . . and verily, for the

righteous, a beautiful place of (final) return." (Qur'an 38:49) [16]
Thus, Adam, Eve, Satan, and the forbidden tree could not have
been in Paradise (Jennat-ul-Khuld), and the notion that the garden
with the forbidden tree and Paradise were one and the same is
false. The Garden of the Forbidden Tree is a place for human trial
and error, while Paradise is a place for reward.

Since the Qur'an does not support the notion that the garden
was Paradise, where then was it? The following verses offer
clues.

After God presented Adam to the Angels, He said to Adam:

> "There is therein (enough provision) for thee not to go
> hungry nor to go naked, nor to suffer from thirst, nor
> from the sun's heat (ضحي)." (Qur'an 20:118-119)[17]

God provided Adam and Eve with food, water, and shelter to
protect them from the heat of the sun. From among the billions of
stars in our universe, why would the Qur'an mention the sun's
heat if the garden was Paradise? Moreover, in every context, the
word ضحي (daha) or its derivatives are used in the Qur'an only
in situations related to the sun. Thus, the Garden of the Forbidden Tree must have been within the solar system in order for it to
be affected by the heat of the sun, which could be felt only on
planets close to it.

The Garden of the Forbidden Tree could not have been on
Mars, Jupiter, Saturn, Uranus, Neptune, or Pluto because their
surface temperatures are so low that mention of the protection
from the sun's heat would be illogical. The surface temperatures
of Mercury and Venus are extremely high, too high for plants and
animals to exist as living creatures and as sources of food for
Adam. The only place in the solar system where DNA-based life
exists and where humans need protection from the sun is Earth.
Based on our present-day knowledge of the solar system and on
the above verses, we can conclude that Adam and Eve were in a
garden that was located on Earth. (Figure 9-1).

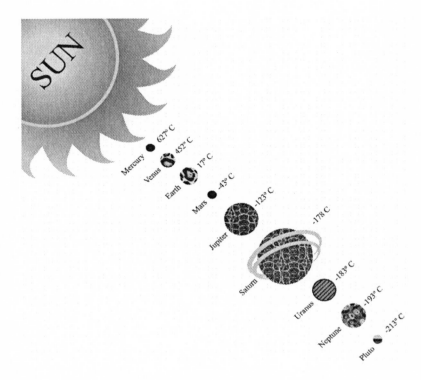

Figure 9-1. SOLAR SYSTEM. Relative sizes of the planets arranged in order of distance from the sun and the surface temperatures of the planets.

The Qur'an also tells us that God created Adam and Eve on the earth with earthly material and granted a fixed term of life there: "Behold, thy Lord said to the Angels: I will create a vicegerent (*khalifah*) on earth" (Qur'an 2:30)[18] and "It is He who created you from clay, and then decreed a stated term [of life] for you . . . ." (Qur'an: 6:2)[19]

Finally, the name bestowed by the Qur'an on the first perfect human being is highly significant. The name "Adam" is derived from two Arabic root words, *adim* and *udama*. *Adim* means surface of the earth, and *udama* means dark-complexioned.[20] The literal meaning of Adam, therefore, is a dark-skinned person who lives on the surface of the earth. This, too, supports the idea that Adam was created on earth and not in Paradise, and was of African origin.

As previously emphasized, the Qur'an asks human beings to observe and discover the mysteries of life on Earth, as well as in the universe:

> On the earth are signs for those who have faith (in the meaningfulness of all things), as also (there are signs) in your own selves: Will you not, then, observe? (Qur'an: 51: 20-21)[21]

What are those mysteries, and what should we observe? The Qur'an is urging us to trace and explore the human evolutionary journey of our primordial parents, Adam and Eve. Thus, we must remain aware of important scientific facts: First, protoplasm is much the same in all organisms, suggesting a unity among all of them. Similar physiological pathways, catalyzed by similar enzymes, occur throughout all of life. At the molecular level of all life forms, we find homologous series of chemical molecules and compounds. The degree of differentiation in the sequence of molecules in the homologous compounds is roughly proportional to the degree of taxonomic separation (i.e., classification of animals or plants according to their natural relationship) of the species in which the compounds occur.

We also need to realize that fossil records document the progression of life through time, and that we can change the shape and function of life forms by changing the genes. We observe that hemoglobin in the red blood cells has two chains of amino acids—alpha and beta chains. Closely related organisms differ less in the amino acid sequences of their hemoglobin alpha chains than do distantly related organisms. The same is true when beta chains are compared with each other.

The above observations suggest that all living forms are related to each other. The grass and the grasshopper and the monkey and humankind can be sisters, brothers, cousins, grandparents, and grandchildren. They all developed from a single seed through genetic modification.

Some Muslims argue that present scientific observations could be proved wrong in the future. Should one reject the Qur'an if further scientific discoveries replace the present paradigm with a new one? Definitely not! Muslims may encounter contradictions between their understanding of what the Qur'an reveals and what they observe in the world. In such cases, it is their duty to point out the flaws in the specific scientific theory with the aid of experimental data.

Many Muslims who believe that Adam and Eve were created in Paradise promote their view by quoting verse 2:38. According to the verse, when Adam disobeyed God's command and ate the forbidden fruit, God ordered him out of the garden, saying: "Get ye down all from here." (Qur'an 2:38)[22] Some Muslims claim that the verse suggests Adam going from a higher place to a lower place. They believe that Paradise is located above the earth.

Such belief is problematic and unacceptable; we know the earth rotates on its axis and is an integral part of the Milky Way Galaxy, which also rotates on its axis. Therefore, if Paradise were visible from Earth, we would see it beneath us at one point in time, above us at another, and at various times on either side. How, then, can we explain the above verse?

Many early Muslim commentators believed that, before their expulsion, Adam and Eve experienced a mental state filled with ease, happiness, and innocence and that the phrase "get ye down" shows they were driven out of a blissful mental state. Muhammad Rashid Rida records that explanation in *Tafsir-ul-Manar.*[23]

Regardless of one's concepts regarding the location of Paradise, based on the above study of the literal reading of the Qur'anic verses, one can reasonably conclude that Adam, Eve, and the forbidden tree were located on the earth and not in the Garden of Eternity.

CHAPTER 10

God and Chance Events

Chance or random events are an integral part of the theory of evolution. What do we mean by chance? The simple meaning of the word is something that happens unpredictably, without discernible human intention or direction, and in dissociation from any observable pattern, causal relation, or natural necessity.

We define events as unpredictable in two ways. In the first, we are limited by the lack of detailed and accurate knowledge at the micro-level. In such cases *chance* means that we are unable to determine all the micro-factors affecting the initial conditions that determine the macro-events. In the second scenario, the observed "unpredictable" event is the result of the crossing of two or more independent causal trains of events, when we have had no precise information about the trains themselves or about their point of intersection, or about the origins of preceding small, sequential, causal events.

Many Muslims believe that assigning a role for chance in the creation of the universe and its subunits—animate and inanimate—is anti-Islamic, a denial of God's absolute control of the universe. In their view, the universe is like a clock and God is the clockmaker. The clockmaker set the clock in motion to tick away toward the Day of Judgment. This miraculous clock never slows down, never runs fast, and never stops. A perfect clock!

But if the universe does run like a clock, then future events should be precisely predictable. Our whole lives and futures would be predetermined, and we would have no freedom to actualize any possibility or affect any change in the universe. If the universe is a mechanical clock whose function was predetermined in the past, all God could do is helplessly observe it ticking away to the Big Crunch. We would have an unemployed God!

A universe without chance and variety of possibilities exists only if creatures do not have the potential or freedom to help or to hurt, to believe or not believe in God, and so on. In a deterministic, chance-absent universe, humans cannot choose freely what they want, but are forced to submit to destiny, and God is responsible for unthinkable evils.

In such a universe, for example, Wahshi, a hired killer, had no choice but to kill Hamzah, the uncle of the Prophet, in the Uhud war, because it was meant to be. Hind d. Utba, who paid Wahshi to kill Hamzah and who then cannibalized him, also had no choice in the matter because it was destined to happen. In such fatalistic universe, Osama bin Laden and other fanatics among the minority of misguided Muslims played no role in forming American public perception of Muslims. Instead, all is destiny! God becomes the ultimate puppeteer who pulls the strings on puppets such as Wahshi and Hind d. Utba to kill and mutilate one of the companions of the Prophet (pbuh), and God becomes responsible for hiding the true face of Islam from ordinary Americans so they see all Muslims as terrorists. In a deterministic, chance-absent universe, human choices and actions are useless because destiny, not chance or volition, transforms them into believers or nonbelievers!

We can discover many other reasons to reject the deterministic, chance-absent universe of Muslim anti-evolutionists.

According to the Qur'an, the current universe is a testing ground for us to gather data in preparation for the Day of Judgment, when God will either reward or punish us based upon His evaluation of what we gathered during our lifetime. The Qur'an states: "[Allah] Who created death and life in order to try you to see who of you are best in deed" (Qur'an 67:2)[1] and "Or do ye think that you shall enter the Garden (of Bliss) without such (trials) as came to those who passed away before you?" (Qur'an 2:214)[2] It also reads:

"On no soul does God place a burden greater than it can bear. It will receive every good it earned and suffer every ill it earned . . . ." (Qur'an 2:286)[3] Therefore, in the Islamic universe, every human being is a free agent, given equal chances and the free will to perform moral or immoral acts—equal chances to get to heaven or hell. What each human being does on this earth matters on the Day of Judgment, when the Divine's absolute justice prevails.

In the predestined universe posited by anti-evolutionists, our thoughts and resulting deeds are preordained by the clockmaker. But if this were so—if Allah prompts every move of His earthly creatures—genuine tests or trials as described in the Qur'an on Earth or Judgment Day are impossible. Why should humankind be accountable for deeds over which they have no control? The Day of Judgment becomes a tyrant's phony court hearing where preprogrammed robots called humans, who have no control over their actions or decisions, are capriciously judged. Without genuine freedom and the chance to do good or evil, where can humankind find the Generous (Al-Karim), Merciful (Ar-Rahman) God?

Many medieval Islamic scholars and imams have rejected a deterministic universe. For example, in response to a letter from Caliph Abd al-Malik ibn Marwan regarding the doctrine of predestination, Imam al-Hasan al-Basri (b.624) replied:

> O Commander of the faithful; do not alter it or interpret it falsely. God would not openly prohibit people from something and destine them to do it secretly as the ignorant and the heedless say. If that were so, He would not have said in the Qur'an, 41:40: *Do what you wish* but would have said: 'Do what I have destined you to do'. Nor would He have said in Qur'an, 18:29: *Whoever wills shall believe and whoever wills shall disbelieve* but would have said: 'Whoever I will shall believe and whoever I will shall disbelieve'.[4]

The fundamental purpose of the Qur'an, and the Prophet's authentic tradition, is to make us aware of heaven and hell and of the moral or immoral decisions that we could make by the use of our free will in an unpredictable world. These sources are also decision-making guides as to what actions God allows, prohibits, or considers neutral. God does not, however, totally determine or preordain individual choices and actions. In this universe, we are free to choose from contrasting possibilities that are packed as information in each moment of the arriving future. We are free to actualize our choices into visible monuments of Allah's creation in the material world.

God holds us accountable if our choices violate the guidance that He revealed though his prophets. All-knowing God (*al-Alim*) knows that free will exists only where authentic choices exist. So He offers us an open future with chance events and a multitude of possibilities that allow us to exercise our free will. Human beings, unlike robots, are even allowed to accept or to reject God, because they are not preprogrammed. No! Allah is not a puppeteer. He is the All Compassionate, Merciful, and Just God, who presents us with possibilities and who permits us the freedom to choose right from wrong or vice versa. Thus we see that random and chance events inevitably occur in this universe peopled with creatures granted with free will.

Anti-evolutionist Muslims acknowledge the truth that Allah used the elements of chance and unpredictability in the historical processes that produced our contemporary universe. Anti-evolutionist Muslims cannot deny this fact, because the revelation of many Qur'anic verses emerged as a direct answer to the free human choices of the Prophet (pbuh) and his contemporaries. A marvelous example is the revelation of chapter 80, "He Frowned (Abasa)." One day the Prophet (peace be upon him) was engrossed in a conversation with some of the most influential chieftains of pagan Mecca, hoping to convince them of the divine message. At that point, a blind ordinary man approached him with the request for an elucidation of certain earlier verses of the Qur'an. Annoyed by this interruption of what he momentarily regarded as a more important endeavor, Muhammad (peace be upon him) "frowned and turned away" from the blind man. Immediately, there and then, the first ten verses of the chapter 80 (He Frowned) was revealed to reprove the Prophet's treatment of the blind man.[5] If chance exists in the daily affairs of human beings and even of the prophet and shapes their histories and epics, even their religions and holy books including the Qur'an, why should Muslims reject the role of chance in shaping our biological history? We can discern no reason to do so in the matter of the evolution of the universe. Chance is an integral, unavoidable part of the evolution of life.

How did God design and direct this miraculous universe yet allow for chance and random events? What are its mechanics? How does God respond to and answer prayers without violating natural laws? How does He maintain the continuity and directionality in the evolution of the universe and life in the presence of chance? Let us explore these questions in the next chapter.

CHAPTER 11

The Structure of the Universe:

A Master Design

Materialists say that the universe is not intelligently designed. Herein, however, we explore the structure of the universe as a product of an intelligent master design to serve the purpose and intent of God in the presence of free will in His creatures. Without understanding the structure of the universe, we cannot understand the creation intelligibly. We begin, therefore, with a review of the basic structure of the universe.

As mentioned earlier, materialistic science claims that the future is totally predetermined by earlier causes and is predictable through accurate knowledge of the past because physical and chemical laws are invariant. Consequently, any event in nature that departs from an anticipated outcome of a known cause is considered coincidental.

Materialists, especially some molecular biologists, arrogantly state that ultimately all aspects of life will be explained in mechanical terms by the knowledge of past events as recorded in geology, paleobiology, vestigial organs, and so on. They conveniently forget a particular past event that shocked materialistic, mechanical science—the quantum theory of Max Planck and its modification by Niels Bohr. Even Bohr was overwhelmed and said, "Anyone who is not shocked by quantum theory has not understood it."[1] In the hands of Niels Bohr, the theory melted solid matter—the basic building block of the universe and its contents (including human)—into nonsolid energy. Twentieth-century relativity and quantum mechanics overturned and swept aside the concept of a predetermined universe and replaced it with an indeterminate universe.

According to the quantum theory, the behavior of matter is unpredictable. No one can say, for example, that a particle was in a definite position in the past, or that it will occupy a definite place at some time in the future.

> A particle such as an electron does not appear to follow a meaningful, well-defined trajectory at all. One moment it is found here, the next there. Not only electrons, but also all known subatomic particles—even whole atoms— cannot be pinned down to a specific motion"[2]

The more accurately we measure the momentum, the less we can calculate the position of a particle. The more we know about the particle's position, the less we can say about its momentum. Our partial information about the position of a particle only yields the probability that it is within a certain distance of a particular point.

The probability factor of quantum mechanics presents itself in other ways. For example, when an atom collides with a photon of light, we cannot precisely predict what will happen, but can only speak of the probability that the energy of the photon will or will not be absorbed by the atom. Similarly, it is not possible to predict when a radioactive atom will decay. In a uranium (^{238}U) sample, the sudden decay of a specific atom into Thorium (^{234}Th) by emitting alpha particles can only be computed as a probable occurrence. We are unable to explain why a particular uranium sample decayed, while the uranium next it with absolutely identical atoms did not. We may calculate a certain chance that the decay will take place within the next ten seconds. The probability exists that the remaining atoms may decay over a period of ten thousand years. But no one can give a definite answer regarding the sequence and time of decay.[3]

Thus we see that the physical sciences cannot tell us when an event will take place, and sometimes it cannot even predict what will happen. If science cannot tell us what is occurring with matter, the basic building block of the universe and living organisms, how can it precisely predict the future of human beings or anything else in the universe? This scientific paradox bothered even Einstein, who nonetheless concluded, "God does not play dice."[4]

The quantum theory gave birth to new chemistry that explained how atoms are bonded together to form molecules. When two atoms initially separated are brought together in a chemical reaction, the electrons in their outermost shell (electrons the farthest from the nucleus) share one orbit. Such sharing in a chemical reaction is called a covalent bond (Figure 11-1).

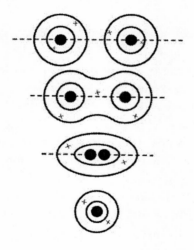

Figure 11-1. FORMATION OF COVALENT BOND.

In covalent bonding, atoms share electrons to form all molecules, including ordinary substances such as water, methane, and so forth. In some cases covalent bonding can lead to the formation of huge, extended macromolecules such as polymers. One example of such a polymer is deoxyribonucleic acid (DNA), the basic building block of life.

The formation and maintenance of the structure of compounds such as DNA is ultimately the result of quantum mechanics, where the behavior of atoms is predictable only as a statistical average. Every thought and every motion of all life forms is inseparably connected to electricity and chemistry, and ultimately with unpredictable quantum events. If the behavior of matter and the formation of compounds are not absolutely predictable, can the future of any creature be precisely predictable?

The reply of quantum physics is a resounding "No" because every DNA-based creature is only a mosaic made of atoms and chemicals. The most that science can calculate are the probabilities of thousands of quantum events as an average. In the quantum world, science loses its confidence in the law of causality. Hence, materialists' extremism against God is exposed as unscientific at the quantum level.[5]

Unlike materialist scientists, who proclaim that there is no purpose for the universe, the world's major religions assert that God created the universe by design and for a purpose. The Prophet Muhammad (pbuh) explains it as follows: "Allah said: I was a hidden treasure. I wanted to be known, so I created the world."[6]

Allah did not want to place all of His creations in the same spiritual reality. Therefore, He created a separate universe. Allah made the larger components of the universe comprehensible to the human mind, the best of His creations, with the installation of stubborn, immutable chemical and physical laws to function in a uniform and repetitive fashion at the level of those larger components. God did not intend the same immutable laws of classical physics and chemistry to work like a programmed machine all the way down to the atomic and subatomic levels.

If the atomic world were functioning in uniform and predictable fashion, then we would be able to foresee the future accurately through our knowledge of past causes. If this were the case, humankind would lose their freedom to make moral choices. Allah would then be restricted to the role of a passive spectator unable to act without suspending the laws of the universe in a way that is visibly obvious to humans.

Therefore God created matter as the building block of the universe and had it governed by an immutable system of physical and chemical laws within which it functions in unpredictable ways. By doing so, He granted human beings the freedom to make moral choices about their future.[7] Due to the unpredictability of the structure of matter, absolute knowledge of the nature and future of our universe will never be known to us until the Day of Judgment.

Materialist scientists are unable to understand that God is active participant in the world. Chemists have discovered that all components of the physical universe, including genes, are made of atoms arranged in different fashions. Genes are made out of DNA. DNA is a collection of nucleotides. Phosphate, sugar, and four amino acids (thymine, cytosine, adenine, and guanine) are chemically connected with covalent chemical bonds to form nucleotides. A gene mutation can produce a major effect on the outward physical features of an organism. Mutation depends on the changes in individual molecules as a result of breaking specific atomic covalent bonds that involve quantum mechanical processes. Physicist-theologian Robert J. Russell points out that "this is ultimately a quantum process at the atomic level initiated by the breaking of a single hydrogen bond."[8]

When God prompts a small quantum fluctuation (jump of a quantum of energy) in the atomic or subatomic world and makes or breaks a covalent chemical bond or bonds of a gene, materialists see a mutation or an accidental birth of a species. They interpret it as a random or accidental event with no cause. They do not realize that God built life around chemistry that provides "the amplifying mechanism for quantum events."[9] Physicist William Pollard feels that if chemistry is the physical appearance of an organism (phenotype), the quantum fluctuation is the cause (genotype).

This construction of the understandable chemistry coupled with the unpredictable quantum physics is an ingenious intelligent design of the All Knowing and All Powerful Allah. Within this master design of the universe, Allah, by voluntarily limiting His omnipotence and omniscience, gave His creatures a genuine freedom and an open future, while at the same time, humans can derive sensible meanings out of the complex universe.

In such a material universe, Allah can, in response to His creatures' prayers, intervene in the universe by causing small quantum events without suspending the understandable laws of classical chemistry and physics. Similarly, in such a design of the universe, God has freedom to create any being or substance, living or nonliving, without disturbing any laws of classical physics and chemistry, by making a quantum event in the atomic world.

In this material universe, therefore, God does play dice between the Big Bang and the Big Crunch by sending messenger moments of the future containing His proposals to His creatures with limited free will.

CHAPTER 12

Self and Subjectivity of the Animate and Inanimate World

Materialists recruit genetic mutation, unexplained events, and pain and suffering in the material world as "proof" to reject God and discredit believers. Within the Islamic metaphysics of the future, however, these natural events are not contradictory evidences against God or against the intelligently designed universe, but complimentary glorification of God.

As we have seen, the creatures in the Islamic universe, which includes quarks and atoms as well as collections of atoms such as the stars, Earth, mountains, human beings, amoeba, etc., are constantly addressed by divine acts that present them with future possibilities, in the form of information.

Allah endowed all creatures with mind. Not only animate but also inanimate creatures, such as fire, wind, and mountains, have self and a subjective faculty with which they can experience and respond to the Divine Will. The Qur'an attests to the existence of subjectivity with faculty to experience and respond within the inanimate world, even though human beings do not comprehend it:

> The seven skies, the earth, and all that lies within them,
> sing hallelujahs to Him. And there is nothing that does
> not chant His praise. But you [human] do not understand

their hymns of praise. Truly He is verily clement and
forgiving. (Qur'an 17: 44)[1]

We also read:

Then he turned to the heavens, and it was smoke. So he
said to the earth and the heavens: Come with willingness
and obedience and they replied: We come Willingly.
(Qur'an 41:11)[2]

The latter verse can be interpreted scientifically. According to
contemporary cosmology, the entire universe was filled with
radiation and plenum of matter (originally hydrogen and helium)
formed from the elementary particles (quarks) in a dense, primeval
fireball of creation called the Big Bang. The smoke, described in
the above verse, is most likely a reference to quarks and atoms
before they condensed into galaxies.

The Qur'anic verses here describe a universe that was responsive
to Allah since its inception after the Big Bang. The heavens and the
earth in its early gaseous embryonic state ("smoke") respond to God
by saying, "We come willingly."

Moreover, the Qur'an substantiates that natural events such
as thunder, fire, and wind have self and subjectivity: "And the
thunder extols His praise, and the angels are in of awe of
Him . . . ." (Qur'an 13:13)[3] The subjectivity of fire is well
documented in the verses. When Abraham was cast into fire,
Allah said: " . . . O fire, be you a coolness and a safety for
Abraham." (Qur'an 21:69)[4] A verse relating to Solomon reads:
"So We subjected the wind to him [Solomon]; it ran softly at
his command to wherever he pleased." (Qur'an 38:36)[5] These
verses guided Jalaluddin Rumi to write: "Air and earth and water
and fire are (His) slaves. With you and me they are dead, but with
thy God they are alive."[6]

According to the Qur'an, Allah offered the heavens, earth, and
the mountains the opportunity to bear the responsibility of carrying

out His will, giving them the choice to deviate from it at the risk of punishment. They refused because they were afraid of violating it and of punishment. The verse reads:

> Surely We presented the trust to the heavens and the earth, and the mountains, but they declined to bear it and they feared it, and man bore it, but he is iniquitous, ignorant. (Qur'an 33:72)[7]

Ibn Kathir (1160-1234), an important medieval exegete of the Qur'an, noted many Traditions of the Prophet (pbuh), from that of Abdullah Ibn Abbas, a companion of the Prophet, to that of Hassan al-Basari (d.728), that reflect the subjectivity of the inanimate world.[8] Similarly, Qurtubi (d.1273), another well-respected exegete, using the authoritative text of Tirmithi (d. 892), who had quoted Abdullah Ibn Abbas, wrote that after the refusal of the heavens, the earth, and the mountains, Allah offered this trust to Adam, who accepted it.[9] Thus, the Islamic faith and its traditions confirm the subjectivity of the inanimate world.

In the spiritual universe of Islam, even a part of a whole organism has its own inherent individual subjective faculty with which to experience and to respond and has a separate physical existence from the whole, as we can see from the following verse:

> This day [Judgment Day] We seal their mouths, and their hands speak to Us and their feet will bear witness concerning what they earned. (Qur'an 36:65)[10]

Another verse warns that on the Day of Judgment:

> . . . their ears and their eyes and their skins will testify against them as to what they used to do. And they say unto their skins: Why ye testify against us? They say: God hath given us speech, who giveth speech to all

things, and who created you at the first, onto whom ye
are returned. (Qur'an 41:20-21)[11]

Thus we see that the individual components of organisms,
such as their DNA, genes, ears, feet, skin, etc., have some measure
of feeling that would allow them to respond to the guidance of
Allah.

The inner potential in lower orders of creatures may be minimal
and hard to detect. As Allah created higher orders of being, He
carefully gifted each of them with more potential, so they could
experience the outside world and be able to understand more
complex messages from Allah. Materialists ridicule the idea of
interiority in inanimate beings. They insist that nature is totally
mindless without providing supporting, experimental data.
However, the famous Young's two-slit experiment, described below,
suggests otherwise.

If we drop two pebbles a few inches apart into a quiet
swimming pool, two wave circles start, each having a top (crest)
and a bottom (trough). At the point where the two crests meet,
the wave is amplified. When crests and troughs from different
waves meet, the waves are canceled out, and the water is not
disturbed. The same phenomenon occurs when light illuminates
a screen with two slits, and it is also true with electrons, gamma
rays, microwaves, and so on.

In an experiment with electrons, a barrier with two slits is
placed in front of a phosphorescent screen such as that of a
television. One electron at a time is shot at the screen. If both slits
are open, the screen will show alternating series of fuzzy dark and
bright diffraction bands (Figure 12-1). If only one slit is open,
there will be one bright fuzzy band (Figure 12-2). If the experiment
is done with each slit closed for half the duration of the experiment,
we see a double exposure separated by a distance that is equal to
that between the slits (Figure 12-3). The patterns on the screen
suggest that the electrons alter their travel patterns depending on
the changing status of the slits. Based on this observation, we

have to assume that the electrons somehow know whether the slit is open or closed.

Another step in the experiment is to determine which slit the electrons passed through. A particle detector is placed around one of the slits to record the path of the particles when the two slits are open and closed for half the duration of the experiment (Figure 12-4). Although both slits are open, we do not see a series of dark and bright interference patterns as we have seen in Figure 12-1. Instead we see the pattern shown in Figure 12-3.

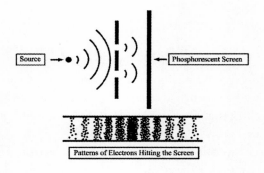

Figure 12-1. BOTH SLITS OPEN.

Figure 12-2. ONE SLIT OPEN.

Figure 12-3. EACH SLIT CLOSED FOR HALF THE TIME.

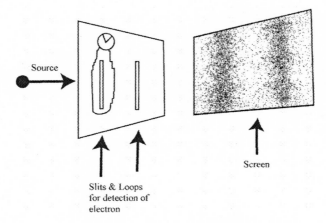

Figure 12-4. BOTH SLITS OPEN, WITH DETECTOR.

The second step of this experiment also suggests that the particles know that both slits are open. They are also able to sense the presence of the particle detector. The changes in their travel patterns confirm that the inanimate beings have mind, as revealed in the Qur'an.

Moreover, no one can explain how an electron jumps from one orbit to another. Ernest Rutherford, a famous physicist, stated that, "an electron would have to 'know' beforehand to which orbit it was going to jump. Otherwise, it would not emit light with a single, definite wavelength when it starts its leap."[12] Recently a few modern Christian scientists, for example Alfred N.Whitehead and John F.Haught[13], have accepted the presence of subjectivity and inner sense within all life forms—from single cell organisms to the most complex human being. Whitehead accepts subjectivity for inanimate beings also.

The Islamic God is a compassionate God who shows mercy for His creatures by voluntarily limiting His power so they can exercise free will to choose good over evil or vice versa. We have seen earlier in verse 10:99 that God has the power to make everyone a believer, but he did not. He commanded the Prophet (peace be upon them) not to coerce people to believe. The Qur'an also states: " . . . [know] that Allah is not a tyrant to His servants"

(Qur'an 8:51)[14] and " . . . Verily Allah never will change the condition of people unless they change it themselves (with their own soul)." (Qur'an 13:11)[15] The Qur'an makes it clear that Allah is a guide, not a tyrant, and that He gives the freedom to His creatures to shape themselves and their future. We have witnessed His compassion when he did not compel the heavens, the earth, and the mountains to accept His offer.

We acknowledge small and large DNA-based creatures as parts of a greater whole called life. Similarly, inanimate objects such as stars and galaxies, the earth, the continents and subcontinents, states and nations, and so on, are atomic aggregates. The collection of elements, such as fire, thunder, tornado, and floods, as well as animate creatures—from genes to viruses, amoebas to humankind—are also atomic aggregates. Environments as tiny as a small culture plate (petri dish) with a growth of bacteria in a laboratory, and as huge as North America or Asia are parts of a whole called nature. The "Gaia hypothesis" of James Lovelock and Lynn Margulis maintains that the biosphere is in some sense a single "organism," a self-regulating whole that operates by its own laws.[16]

I believe that it operates by Allah's laws. Based on the Qur'an, we must expand the definition of biosphere to include the universe as a whole, under whose umbrella all individual units and subunits, or all separately identifiable units, are seen as organisms with subjectivities and selves.

Creatures can act within certain boundaries that Allah has specifically ordained for each subunit. For example, the electropositive ions potassium (K^+) and sodium (Na^+) cannot bond together to make a compound, while potassium or sodium is free to bond together with an electronegative ion, chloride (Cl^-).

Allah also created varying levels of inner sense and subjectivity with faculty to experience and respond for whole systems and for their parts. Therefore when He addresses His creatures, He mercifully expects different levels of response from them. We have seen earlier that inanimate units such as fire, thunder, etc., have duties and functions. Even "seventh heaven" has certain duties, as stated in the verse, " . . . and He assigned to each heaven its duties and command . . . ." (Qur'an 41:12) [17]

When nature or any of its parts responds to possibilities (proposals) in the arriving moments of future from God, we interpret its actions in terms of natural selection. We observe nature deselecting the unfit choices of its own parts. In chapter 6 we saw Ibn Khaldun and al-Biruni attributing functions and duties to nature.

The complexity of information or future possibilities arriving for the use of any hierarchical order of being depends upon that order's inherent level of intellect and capability to actualize the information. Obviously, Allah did not want an amoeba to fly an airplane, but He wants human beings and *Jinn* to explore the heavens, as clearly revealed in the following verse:

> O company of *jinn* and men, were you able to penetrate
> bounds of heavens and the earth, do then penetrate
> (them). Yet, you shall not pass through save by an
> authority. (Qur'an 55:33)[18]

Creation/evolution of life is the result of Allah presenting possibilities (proposals) in each arriving moment of the future to the atoms as well as the aggregates of atoms. Within boundaries set by God, creatures, animate and inanimate, are free to actualize these possibilities into visible monuments of divine creation.

In the Islamic universe, all possibilities—pleasant or unpleasant—are from God. The Qur'an states, "No misfortune befalls but by God's leave." (Qur'an 64:11)[19] However, the actualization of good and bad from the arriving future belong to His creatures. The free choices of the whole and its subunits make our universe. In this unpredictable universe with an open future, creatures can get hurt, but they can prosper, too. The pain and suffering of the components of the material universe is the price we pay for our free will. It is a place where we can experience life, but will also be tested for the veracity of our faith in Allah through adversities, as the verse states:

> And We shall test you with something of fear and hunger
> and loss of wealth, and lives and crops. Yet give good
> tidings to the patient. (Qur'an 2:155-157)[20]

We have to acknowledge that evolution is a fact. The evolution or unfolding of the universe is the process of creation by Allah. It did not come into being solely by coincidental causes in the past that create the future, as materialists claim. The physical universe and life evolved due to enterprising creatures, armed with past experiences, choosing one or more possibilities within each moment flowing from the future but intended by Allah.

Those lazy creatures who did not respond to the choices arriving from Allah through the messenger moments of the future remained what they are. Therefore, I feel no shame in accepting what I am as a result of my humanlike ancestors and the pre-human tailed creatures who preceded them and accepted Allah's offer to make them humans when He told the angels, " . . . I am setting on earth a vicegerent . . . ." (Qur'an 2:30)[21] In fact, I am proud of my pre-human ancestors' genes, which chose to receive and grasp Allah's guidance to help our ancestors to transform into humankind. Of course, the evolution of life into various species occurred as a result of His mercy to limit His omnipotence and omniscience voluntarily. Therefore, evolution is a fact of life as a result of Allah's gift of freedom to His creatures.

Ian G. Barbour, physicist and Christian theologian, describes the process of creation/evolution splendidly in his book, *Religion and Science.* He compares God to a Choreographer of a dance in which the dancers are free to make many decisions by themselves.[22] In the creation/evolution of the Islamic universe, we dancers and musicians do not invent the better/worse or expanded/abridged versions the dance or the symphony. By our own choice we only objectify in our time and space any one of the variations of dance or symphony contained in the messenger moments of the arriving future from God. This analogy of God as choreographer or composer also clarifies the Islamic doctrine that goodness and evil are from God and humans actualize it by their choice.

Within the Islamic metaphysics of the future, the Islamic universe is always under Allah's providence and never leaves His hand because He has the absolute power and free will to decide what possibilities to offer us. Human as well as all other separately identifiable subunits of the universe have the total freedom based upon individual competency to choose and actualize the offered possibilities into visible monuments of Allah's creation. Just as for humankind, Allah has given a variety of freedoms to all subunits as well as the whole universe. Similar to the way of humanity, every subunit accepts or rejects God's guidance to choose from possibilities presented to them. All units freely select evil or righteous possibilities to get some benefit in this world or in the Hereafter Universe (*al-Akhiram*). In such an unpredictable universe, some get hurt while others do not. (Please refer to chapter 6 for ample evidence from Jalaluddin Rumi to support this understanding.)

Based upon the teachings of the Qur'an, the far-reaching meaning of the phrase *Inshah Allah* (God so willing), and the concept of Islamic metaphysics of future, we see that, while evolutionary accidents or contingencies appear random, they come from Allah as novel suggestions. The pivotal moment called "present" is the opening into the material world.

When messenger moments with novel possibilities or potentialities arrive at this door and when a creature objectifies it, we may see a mutation or the birth of a new species. An event in nature may appear random to humans, because we have no knowledge of what the messenger moment of the future has brought to every creature in the universe, and we also have no means to know what every individual being (separately identifiable components of the universe) has chosen to actualize into monuments of divine creation. Globally the pain, suffering, or bliss that creatures experience are the normal outcome of the free choices creatures make among the possibilities they encounter at their individual levels when the future bursts into the razor-sharp edge of a fleeting present moment.

By granting freedom to His creatures to choose from the possibilities contained in the arriving messenger moments of future, Allah distanced Himself from being a tyrant and distinguished Himself as the All Merciful (*ar-Rahman*) and The Evolver (*al-Baari*). He only proposes and creatures dispose.

Finally, Allah is the source of all possibilities contained in the arriving future. Therefore, He becomes the Originator (*al-Baadi*), the Shaper (*al-Musawwir*), and the Evolver (*al-Baari*) of all things. The final result is that we live in a dynamic universe that is alive and being created in every moment in the dazzling dance between Allah's limited creatures and His unlimited imagination.

SUMMARY

Although many contemporary Muslims believe that God instantaneously created a well-designed world, science has shown that the universe and life on earth evolved over billions of years. Similarly, the Muslims in their classical period and in the Qur'an believed that creation is a process that occurred over long period of earthly time.

According to the Qur'an and medieval Muslims, all separately identifiable small or large components of the universe have self and subjectivity. All inanimate material beings such as wind, fire, mountains etc., are living creatures. Our evolving universe is the result of the struggle between the small and the large components of the universe in which creatures fight each other to get some benefit and avoid some injury.

As such, the material world is not a perfect place of joy and jubilation. Our material world is characterized by the coexistence of evil and good; pain and joy. God intended it to be a fierce war zone, a testing ground for the evolution of His moral agents and a prelude to their entry to Paradise. The Qur'an describes that, contrary to the imperfection of the transient material universe, the Hereafter world is a place devoid of evil, in which believers enjoy perfect peace: "And surely the world to come is greater in ranks, greater in preferment" (Qur'an 17:21)[1], and "Therein they shall hear no idle talk, no cause of sin, only the saying 'Peace, Peace!'" (Qur'an 56:25)[2]

Application of John Haught's novel and liberating concepts of "metaphysics of future" within the Islamic context unravels a logical solution to the conflict between science and contemporary Islam. It helps to understand the relationship between God's omniscience and human free-will also. I have explained that the expression Inshah Allah (God so willing) represents the Islamic metaphysics of the future. Messenger-moments of future do not exist and never arrive for material beings until they emanate from God. They hold within them information that embodies potential possibilities or proposals that God grants to His creatures. These possibilities remain dormant until a living being, exercising free will at a given moment, actualizes them within a material medium. Thus, our evolving, short-lived, and imperfect material world results from the materialization of chosen possibilities or proposals coming from God. Yet, stunningly, in this potentially chaotic universe, which evolves through the practice of free choice, order emerges. Such order results from the interplay between the organization of possibilities that God makes within messenger-moments and the choices creatures make, based on their past experiences and self-interest, before actualizing the possibilities visibly.[3]

I have argued that, according to the Qur'an, all separately identifiable beings have self and subjectivity. The creation is a process where messenger-moments, holding novel or routine potential for the future, arrive at the brief, living present of all creatures, allowing them to actualize the potential to reality. When potentials become visible realities of the material world, the active present of creatures is kicked out to become their "fixed past."[4] Deterministic scientists, however, believe that the "fixed past" created the universe.

God offers a variety of proposals through messenger-moments. While Allah addresses the lower forms of being with the least complex and limited number of proposals, humans are presented with numerous, more complex possibilities. Each of us has the freedom to choose and objectify any of the proposals in the material of our world. Creation/Evolution is the end-result of the interplay between God reaching out "to the entire universe"[5] with His proposals and His creatures exercising their limited free will to select and transform any of the proposals into visible realities of the material world. New species are born when previously existing beings truly grasp and actualize the novel proposals (potentials) enclosed within imminent moments delivered by God.

The existence of evil or goodness within this world also comes from the choice made by the big and small creatures. Sometimes, choices made by creatures may totally block other creatures' freedom to make choices, leading to frustration for those who reject God. However, when all possibilities are blocked, resulting in calamity or even death, the believing creatures "seek help in patience and prayer," (Qur'an 2:153)[6] and say, "surely we belong to God, and to Him we return" (Qur'an 2:156)[7]. In the defeat or submission during "the war in nature," true believers know that there exists "blessing and mercy from their Lord" for the patient among His creatures (Qur'an 2:157)[8].

The theory of evolution asserts that all life forms evolved from a common ancestor. Organisms that exist today are those that nature allowed to survive because of their adaptability to changing environmental circumstances. Materialists among the evolutionists contend that a random accident called mutation occurs in genes of some species, resulting in variations in species. Materialists argue that the contingent, accidental nature of these evolutionary variations is clear evidence against the Divine.

The theory of evolution is no way anti-Islamic, but such an atheistic and antireligious interpretation of evolution has most probably led to contemporary Muslim clerics' (imams) and scholars' rejection of science. This rejection led in turn to a decline in scientific progress and subsequent underdevelopment in the Muslim world. In mosques, imams preach that Adam and Eve were created in Paradise and expelled from it because the nuclear grandparents chose to trust grand Satan (Iblis) more than God. However, we have learned that the Qur'an and medieval Muslim scholars explained the concept of evolution centuries before Charles Darwin. Historically, Muslims were the first to record the discovery of human origin from the world of apes. Therefore, Adam and Eve could not have been created *ex nihilo* in Paradise, but evolved on the earth.

Allah created the universe with classical laws of physics and chemistry to work at the macro-level of being, while quantum mechanics and indeterminacy rule at the micro—atomic and subatomic—levels. By such a design God maintains His omnipotence to respond to His creatures' needs. The generation of small quantum fluctuations in the micro-world can make major changes in the macro-world without violating the laws of classical physics and chemistry. Materialists, however, interpret God's actions at the quantum level as accidental genetic mutations.

The theology I have discussed blends the Qur'anic revelation, medieval Muslim understanding of nature, Whitehead's process thoughts, William Pollard's quantum physics theology, and John Haught's metaphysics of the future. Evidence for the story of the material universe supports an abrupt beginning at the Big Bang, then an evolution, then a final Big Crunch. God knows all possible numerous directions and roads that this universe and its component parts would take during their journey from Big Bang to Big Crunch, but He voluntarily and mercifully limited His omniscience and omnipotence in order to create freedom for His creatures to choose freely from possibilities that He presents to them with every approaching moment.

The design of the universe, with the indeterminate nature of quantum behavior of basic matter along with the blended theological interpretations explicated herein, allows us to conclude that Allah maintains His omnipotence, omniscience, love, and mercifulness without visibly violating any immutable laws of nature, allowing living beings to enjoy the gift of freedom.

NOTES

NOTES TO INTRODUCTION

1   Moyers, Bill. *A World of Ideas,* p. 272.

2   Yahya, Harun. *The Evolution Deceit.*

3   Ali, Yusuf. *The Holy Qur'an.*

4   Nasr, Seyyed Hussein. *Islamic Spirituality Foundation,* p. xxi.

5   Ibid. p. xxii.

6   Ali, Ahmed. *Al-Qur'an.*

7   Pickthall, Mohammed Marmaduke. *Holy Qur'an.*

8   Ali, Yusuf. *The Holy Qur'an.*

9   Khatib, Dr. M. M. *The Bounteous Koran.*

10   Ali, Yusuf. *The Holy Qur'an.*

11   Arberry, A. J. *The Koran Interpreted,* 1955.

12   Ali, Yusuf. *The Holy Qur'an.*

13   Ali, Ahmed. *Al-Qur'an.*

14   Ali, Yusuf. *The Holy Qur'an.*

NOTES TO CHAPTER 1

1   Barbour, Ian G. *Religion and Science,* p. 81.

2   Richard Dawkins. *The Blind Watchmaker.*

3   Richard Dawkins. *River out of Eden.* New York: Basic Books, 1995.

4   From Richard Lewontin's review of Carl Sagan's book, *The Demon-Haunted World: Science as a Cradle in the Dark,* in the *New York Review of Books,* January 9, 1997.

5   John F. Haught. *God After Darwin.* Boulder: Westview Press, 1999, p. 86.

6   Ibid.

7   Khatib, Dr. M. M. *The Bounteous Koran.*

8   Ibid.

9   Ibid.

10   Ibid.

11   Ibid.

12   John F. Haught. *God After Darwin*, p. 83-88.

13   Khatib, Dr. M. M. *The Bounteous Koran.*

14   Ali, Ahmed. *Al-Qur'an.*

15   Khatib, Dr. M. M. *The Bounteous Koran.*

16   Ibid.

17   Guillanme, A. *The Life of Muhammad,* p. 375.

18   Ali, Ahmed. *Al-Qur'an.*

19   Barbour, Ian G. *Religion and Science,* p. 314.

20   Pickthall, Mohammed Marmaduke. *Holy Qur'an.*

NOTES TO CHAPTER 2

1   Nasr, Seyyed Hossein. *"Islam and the Environmental Crisis,"* MAAS Journal of Islamic Science. p. 35.

2   Eaton, C. G. *Islam and the Destiny of Man,* p. 87.

3   Arberry. A. J. 1955. *The Koran Interpreted.*

4   Khatib, Dr. M. M. *The Bounteous Koran.*

5   Ali, Yusuf. *Holy Qur'an.*

6   Irving, T. B. *The Qur'an.*

7   Al-Ghazzali, Imam Abu-Hamid. (1058-1119 AD). *Ihya Ulum-Id-Din,* p. 20.

8   Khatib, Dr. M. M. *The Bounteous Koran.*

9   Al-Suhrawardy, Allama Sir Abdullah Al-Mamun. *The Sayings of Muhammad,* p. 108.

10   Ibid. p. 114.

11   Ibn al-Hajjaj, Abul Husain Muslim (d.875). *Sahih Muslim,* trans. Abdul Hamid Siddiqui. Book 18, No.4261.

12   Khatib, Dr. M. M. *The Bounteous Koran.*

13   Bukhari, Muhammad ibn Ismail al-Jufi (d.897). *Sahih Al-Bukhari,* Vol. VI, p. 386.

14   Morris, Richard. *Time's Arrows,* p. 10.

15   *Encyclopaedia Britannica, Macropaedia,* "Time," Vol. 18, p. 411.

16   Ali, Yusuf. *Holy Qur'an.*

17   Morris, Richard. *Time's Arrows,* p. 66-67.

18 Hawking, Stephen W. *A Brief History of Time,* p. 7.

19 Morris, Richard. *Time's Arrows,* p. 27.

20 Ibid. p. 209.

21 Davies, Paul. *God and the New Physics,* p. 120.

22 March, Robert H. *Physics for Poets,* p. 124.

23 Sagan, Carl. *Cosmic Connections,* p. 246.

24 March, Robert H. *Physics for Poets,* p. 125.

25 Morris, Richard. *Time's Arrows,* p. 178.

26 Ibid. p.180-181.

27 Asad, Muhammad. *The Message of the Qur'an,* p. 513.

28 Ali, Ahmed. *Al-Qur'an.*

29 Khatib, Dr. M. M. *The Bounteous Koran.*

30 Ibid.

31 Ali, Yusuf. *The Holy Qur'an.*

32 Maududi, Abul A'la S. *The Meaning of the Qur'an.*

33 Asad, Muhammad. *The Message of the Qur'an.*

34 Ibid.

35 Ibid.

36 Khatib, Dr. M. M. *The Bounteous Koran.*

37 Schroeder, Gerald L. *The Science of God,* p. 164.

38 Ibid. p. 164

39 Ibid. p. 164.

40 Ali, Ahmed. *Al-Qur'an.*

41 Khatib, Dr. M. M. *The Bounteous Koran.*

42 Davies, Paul. *God and the New Physics,* p.128.

43 Hakim, Abdul Hakim. *The Metaphysics of Rumi,* p 18.

44 Whinfield, E. H. *Teachings of Rumi,* Inside cover page.

45 Ali, Yusuf. *Holy Qur'an.*

46 Khatib, Dr. M. M. *The Bounteous Koran.*

NOTES TO CHAPTER 3

1 Hardy, Edward R. "Origin and Evolution of Universe." *Encyclopaedia Britannica,* Macropaedia, Vol. 18, p 1007

2 Adler, Mortimer J. *Aristotle for Everybody,* p. 187.

3 Eve, Raymond A., and Harrold B. Francis. *The Creationist Movement in Modern America,* p. 13-14

4   Ibid. p. 48.

5   Ibid. p. 48.

6   Ibid. p. 47-48.

7   Ibid. p. 46.

8   Dods, D.D, Marcus, and others. "Book of Genesis" in *An Exposition of the Bible*, Vol. 1. p. 5-7.

9   Asimov, Isaac. *Beginnings: The Story of the Origins—Of Mankind, Life, the Earth, the Universe,* p. 253.

10   Hardy, Edward R. "Origin and Evolution of Universe." *Encyclopaedia Britannica,* Macropaedia, Vol. 18, p 1007

11   Abell, George O. "The Age of the Earth and the Universe" in *Scientists Confront Creationism,* Godfrey, Laurie, ed., p. 46

12   Sagan, Carl. *Cosmos,* p. 246.

13   Ferris, Timothy (narrator). 1990. *The Creation of the Universe.* PBS Home Video, Beverly Hills.

14   Hawking, Stephen W. *A Brief History of Time,* p. 70-72.

15   Barrow, John D., and Silk, Joseph. 1983. *The Left Hand of Creation,* p. 73-75

16   Hawking, Stephen W. *A Brief History of Time,* p. 70-72

17   Ferris, Timothy (narrator). 1990. *The Creation of the Universe.* PBS Home Video, Beverly Hills.

18   Asimov, Isaac. *Beginnings: The Story of the Origins—Of Mankind, Life, the Earth, the Universe,* p. 263-264.

19   Ali, Ahmed. *Al-Qur'an.*

20   Asad, Muhammand. *The Message of The Qur'an.*

21   Sagan, Carl. *Cosmos,* p. 246.

22   Asad, Muhammand, *The Message of The Qur'an,* p. 491.

23   Ibid.

24   Ali, Ahmed. *Al-Qur'an.*

25   Lane, Edward William. *An Arabic-English Lexicon,* part 2, p. 816.

26   Sagan, Carl. *Cosmos,* p. 246.

27   Davies, Paul. *God and the New Physics,* p. 204.

28   Jastrow, Robert. *God and the Astronomers,* p. 129.

29   Schroeder, Gerald L. 1990. *Genesis and Big Bang,* p. 81.

30   *Did Man Get Here by Evolution or Creation.* New York: Watch Tower Bible & Tract Society of Pennsylvania. p. 186.

31   Ali, Yusuf A. *The Holy Quran.*
32   Arberry, Arthur J. *The Koran Interpreted.*
33   Hardy, Edward R. "Origin and Evolution of Universe." *Encyclopaedia Britannica,* Macropaedia, Vol. 18, p. 1011.
34   Davies, Paul. *God and the New Physics,* p. 204.
35   Arberry, Arthur J. *The Koran Interpreted.*
36   Davies, Paul. *God and the New Physics,* p. 204.
37   Ali, Ahmed. *Al-Qur'an.*
38   Sagan, Carl. *Cosmos,* p. 231.
39   Arberry, Arthur. J. *The Koran Interpreted.*
40   Sagan, Carl. *Cosmos,* p. 231.
41   Arberry, Arthur J. *The Koran Interpreted.*
42   Pickthall, Mohammed Marmaduke. *Holy Qur'an.*
43   Sagan, Carl. *Cosmos,* p. 231.
44   Ali, Yusuf. *The Holy Quran.*
45   Lane, Edward. *An Arabic-English Lexicon,* part 4, p. 1308.
46   Sagan, Carl. *Cosmos,* p. 231.
47   Arberry, Arthur J. *The Koran Interpreted.*
48   Davies, Paul. *God and the New Physics,* p. 204.
49   Salahi, Adil M. and Ashur A. Shamis. 1979. *In the Shade of the Qur'an.*
50   Hardy, Edward R. "Origin and Evolution of Universe." *Encyclopaedia Britannica,* Macropaedia, Vol. 18, p. 1011.
51   Ali, Yusuf. *The Holy Quran.*
52   Ibid.
53   Ibid.
54   Ibid.
55   Ibid.
56   Ibid.
57   Kaufman III, William J. *Black Holes and Warped Spacetime,* p. 84.
58   Hawking, Stephen W. *A Brief History of Time,* p 133.
59   Hardy, Edward R. "Origin and Evolution of Universe." *Encyclopaedia Britannica,* Macropaedia, Vol. 18, p. 1011.
60   Sagan, Carl. *Cosmos,* p. 267.
61   Arberry, Arthur J. *The Koran Interpreted.*
62   Dawood, N. J. *The Koran.*

NOTES TO CHAPTER 4

1   Edey, Maitland A. and Donald C. Johanson. *Blue Print*, pp. 34, 291.

2   al-Faruqi, I. R. *The Cultural Atlas of Islam*, p. 313.

3   Bronowski, Jacob. *The Ascent of Man*, p. 166.

4   Briffault, Robert. 1980. *The Making of Mankind*, pp. 200-201.

5   al-Biruni, Abu Raihan. 1000 A.D. *Athar-ul-Bakiya (The Chronology of Ancient Nations)*. Edward C. Sachu, trans., p. 85.

6   Schroeder, Gerald L. *Genesis and The Big Bang*. p. 31.

7   Asimov, Isaac. *Beginnings: The Story of the Origins—Of Mankind, Life, the Earth, the Universe*, pp. 22-23.

8   Edey, Maitland A. and Donald C. Johanson. *Blue Print*, p. 8.

9   Eve, Raymond A. and Francis B. Harrold. *The Creationist Movement in Modern America*, p. 46.

10 Asimov, Isaac. *Beginnings: The Story of the Origins—Of Mankind, Life, the Earth, the Universe*, p. 26.

11 al-Biruni, Abu Raihan. 1000 A.D. *Athar-ul-Bakiya (The Chronology of Ancient Nations)*. Edward C. Sachu, trans., p. 16-22.

12 Eve, Raymond A. and Francis B. Harrold. *The Creationist Movement in Modern America*, p. 4.

13 Trefil, James. *Reading of the Mind of God*, p.135.

14 Abell, George O. "The Age of the Earth and the Universe," in *Scientists Confront Creationism*. Godfrey, Laurie, ed., p. 35

15 Ibid. p. 43.

16 Marvin, Ursula B. "Continental Drift," *Encyclopaedia Britannica, Macropaedia*, (15th ed.) Vol. 5. p 108.

17 Abell, George O. "The Age of the Earth and the Universe," in *Scientists Confront Creationism*, Godfrey, Laurie, ed., p. 36.

18 Ali, Yusuf. *The Holy Qur'an*.

19 Ibid.

20 *Encyclopaedia Britannica*, Micropaedia, (15th ed.) "Al-Biruni," Vol. II. p. 42.

21 al-Biruni. *Kitab Tahdid al-Amakin Listashiah Masafat al-Masakin (The Determination of the Coordinates of Positions for Correction of Distances between Cities)*. Jamil Ali, trans., pp.14-16.

22 al-Biruni, Abu Raihan. 1000 AD. *Athar-ul-Bakiya (The Chronology of Ancient Nations)*. Edward C. Sachu, trans., p. 30

23   Ali, Ameer. *The Spirit of Islam*, p. 431-432.

24   Plott, John C. *The Global History of Philosophy*, pp. 258-259.

25   Nasr, Sayyed Hossein. *An Introduction to Islamic Cosmological Doctrines*, p. 80.

26   *Encyclopaedia Britannica*, Micropaedia, "Avicenna," Vol. I. p.661.

27   Osborn,H. F. *From Greeks to Darwin*, p. 76.

28   Dodson, Edward O. and Peter Dodson. *Evolution: Process Product*, p. 79.

29   Shadewald, Robert J. "The Evolution of Bible-science," in *Scientists Confront Creationism*. Godfrey, Laurie, ed., 286.

30   Gould, Stephen Jay. *Time's Arrow Time's Cycle*, pp. 119-125.

31   al-Biruni. *Fi Tahqiq Ma Li'l-Hind (Alberuni's India)*. Edward C Sachau, trans. Vol. I, p. 198.

32   Nasr, Sayyed Hossein. *An Introduction to Islamic Cosmological Doctrines*, p. 141.

33   al-Biruni. *Fi Tahqiq Ma Li'l-Hind (Alberuni's India)*. Edward C. Sachau, trans. Vol. I, pp. 378-379.

34   al-Biruni, Abu Raihan. 1000AD. *Athar-ul-Bakiya (The Chronology of Ancient Nations)*. Edward C. Sachu, trans., p. 27-28.

35   Nasr, Sayyed Hossein. *An Introduction to Islamic Cosmological Doctrines*, pp. 80-81.

36   Ibid. p. 245.

37   Ibn-Khaldun, Abdul-Rahman. *The Muqaddimah*. Franz Rosenthal, trans. Abr.ed.

38   Ibid.

39   Ibn-Khaldun, Abdul-Rahman. *Muqaddimah*. Franz Rosenthal, trans. Vol. I, pp. 169-170.

40   Ibid. p. 173

41   Edey, Maitland A and Donald C. Johanson. *Blue Print*, p. 8.

42   Asimov, Isaac. *Beginnings: The Story of the Origins—Of Mankind, Life, the Earth, the Universe*, pp. 46-48.

43   Nasr, Sayyed Hossein. *An Introduction to Islamic Cosmological Doctrines*, p. 142.

44   Osborn, H. F. *From Greeks to Darwin*. p. 76.

45   Draper, John William. *The Intellectual Development of Europe*, p. 42.

46   Briffault, Robert. *The Making of Mankind*, p. 188-190.

47   Ibid 190.

NOTES TO CHAPTER 5

1   Ashley, Montagu (ed.), *Science and Creationism,* "Evolution as Fact and Theory," by Stephen Jay Gould, pp. 118-119.

2   Curtis, Helena and Sue N. Barnes. *Biology,* (5th ed), p. 974

3   Shapiro, Robert. "Probing of the Origin of Life," *Encyclopaedia Britannica: 1984 Year Book of Science and Future,* p. 9

4   Ibid. p.14.

5   Miller, Stanley L., and Robert G. Orgel. *The Origin of Life on the Earth,* p. 33.

6   Shapiro, Robert. "Probing of the Origin of Life," *Encyclopaedia Britannica: 1984 Year Book of Science and Future,* p. 14-15.

7   Abelson, Philip H. "Chemical Events on the Primitive Earth," *Proceedings of National Academy of Science,* Vol.55, p.1365.

8   Dimroth, Erich, and Michael M. Kimberley. "Pre-Cambrian atmospheric oxygen: Evidence in the sedimentary distributions of Carbon, Sulfur, Uranium, and Iron," *Canadian Journal of Earth Sciences,* Vol. 13, September 1976, p. 1161.

9   Henderson-Sellers, A. Benlow and A. J. Meadows. "The Early Atmospheres of Terrestrial Planets," *Quarterly Journal of Royal Astronomical Society.* Vol. 21 (1980), p. 81.

10   Shapiro, Robert. *Origins,* p. 112.

11   Doolittle, Russell F. "Probability and the Origin of Life," *Scientists Confront Creationism,* Godfrey, Laurie R., ed., p. 89-90.

12   Shapiro, Robert. "Probing of the Origin of Life," *Encyclopaedia Britannica: 1984 Year Book of Science and Future,* p. 22.

13   Sagan, Carl E. "Life," *Encyclopaedia Britannica* 15th ed., p. 902. [

14   Shapiro, Robert. *Origins,* p. 164.

15   Edey, Maitland A. and Donald C. Johanson. *Blue Prints,* p. 288-289.

16   Darwin, Charles. *The Origin of Species,* pp. 73-74.

17   Ibid., p. 450.

18   Dodson, Edward O. and Peter Dodson. *Evolution Process and Products,* p. 189-190.

19   Ibid., p 233.

20   Gould, Stephen Jay. *Eight Little Piggies,* pp. 396-397.

21   Shermer, Michael. "25 Creationists' Arguments & 25 Evolutionists' Answers," *Skeptic,* Vol.2, No.2.

22 Futuyma, Douglas J. "World Without Design." *Natural History,* March 1987, p. 34.

23 Dodson, Edward O. and Peter Dodson. *Evolution Process and Products,* p. 543.

24 Mayr, Ernst, and William Provine, eds. *The Evolutionary Synthesis,* "Prologue: Some Thought on the History of the Evolutionary Synthesis," p. 1.

25 Darwin, Charles. *The Origin of Species,* p. 312.

26 Stanley, Steven M. *Macroevolution Pattern and Process,* p. 39.

27 Raup, David M. "Conflict between Darwin and Paleontology," Field Museum of *Natural History Bulletin,* p.25.

28 Gould, Stephen Jay. "Evolution, As Fact And Theory," *Science and Creationism,* Montagu, Ashley, ed., p.123.

29 Gould, Stephen Jay. "A quahog is a quahog," *Natural History,* August-September 1979, p. 18-26.

30 Futuyma Douglas J. "The Theory of Evolution," *Encyclopaedia Britannica* (15th ed.), p. 1005.

31 Dodson, Edward O. and Peter Dodson. *Evolution Process and Products,* p. 178.

32 Gould, Stephen Jay and Nils Eldredge. "Puctuated Equilibrium Comes of Age," *Nature* (1993), p. 223.

33 Futuyma, Douglas J. "The Theory of Evolution," *Encyclopaedia Britannica* (15th ed.), p. 986.

34 Ibid. p.987.

35 Ibid.

36 Dodson, Edward O. and Peter Dodson. *Evolution Process and Products,* p. 52-53.

NOTES TO CHAPTER 6

1 Blackmore, Vernon., Page, Andrew. *Evolution the Great Debate,* Illinois: Lion Publishing Company, 1989, p. 106.

2 Ibid. 102

3 Gould, Stephen Jay. "Knight Takes Bishop?" *Natural History,* (May 1986), p. 18.

4 Lucas, J. R. *Historical Journal,* XXII (1979), p. 102

5 Draper, John William. *The Conflict between Religion and Science,* pp. 187-188.

6 Briffault, Dr.Robert. *The Making of Mankind,* p. 201.

7   Godfrey, Laurie R (ed.). *Scientists Confront Creationism,* "Scopes and Beyond: Antievolution and American Culture," Cole, John R, p. 25-27.

8   Draper, John William. *The Conflict between Religion and Science.* p. 118.

9   Mayr, Ernst. *One Long Argument: Charles Darwin and the Genesis of Modern Evolutionary Thought,* p. 3.

10   Miller, Richard. *The Encyclopedia of Evolution,* p. 116.

11   Hitti, Philip K. *History of Arabs,* p. 588.

12   Draper, William John. *The Intellectual Development of Europe,* Vol. 2, pp. 39-40.

13   Durant, Will. *The Story of Civilization,* Vol. 4, p. 246-249.

14   Tufail, Abu Bakr Muhammad. *Hai bin Yaqzan (The Journey of The Soul),* pp. vi-vii.

15   Miller, Richard. *The Encyclopedia of Evolution,* p. 470.

16   Durant, *The Story of Civilization,* Vol. 4, p. 243.

17   Ibid. 249.

18   Khaldun, Ibn. *The Muqaddimah,* trans. by Franz Rosenthal, Vol. 1, p. 195.

19   Ibid. Vol. 1, p. 173.

20   Ibid. Vol. 3. p. 238.

21   Ibid. Vol. 2, p. 422-423.

22   Hussani, S.A.Q. *The Pantheistic Monism of Ibn Arabi,* p. 61.

23   Ibid. p. 66.

24   Ibid p. 107.

25   Ibid. p. 106.

26   Ibid. p. 61

27   Ibid. p.105

28   Ibid. p. 105.

29   Gould, Stephen Jay. *Ever Since Darwin,* p. 60-61.

30   Biruni, Al (1000 A.D) *The Athar-Ul-Bakiya (The Chronology of Ancient Nations),* translated by Dr. C. Edward Sachu, p.98.

31   Ikhwan al-Safa. *Rasa'il Ikhwan al-Safa wa-Khull an al-Wafa (Epistles of Brothrens of Purity),* Vol. II, p. 178.

32   Ibid. Vol. II, p. 181.

33   Ibid. Vol. II, p. 182.

34   Ibid. Vol. II, p. 192.

35   Nicholson, Reynold A. *Rumi,* p. 11.

36 Hakim, Dr Khalifa Abdul. *The Metaphysics of Rumi*, p. 36.

37 Ibid. p. 36

38 Ibid. p. 37.

39 Al-Biruni. *Kitab al-Jamahir fi ma'rifat al-jawahir, (Arabic)*, Hyderabad: Dairatu'l Ma'arif'l Osmania, 1934, p. 80.

40 Tufail, Abu Bakr Muhammad. *Hai bin Yaqzan (The Journey of The Soul)*, pp. vi-vii.

41 Osborn.H.F. *From Greeks to Darwin*, pp. 77-78.

42 Shadewald, Robert. *National Center For ScienceEducation Reports*, Vol. 10 (March-April 1990), p. 21.

43 *Encyclopaedia Britannica*, (15th ed.) Micropaedia. "Ibn Miskawayah." Vol. V, p. 272.

44 Iqbal, Sir Mohammad. *The Reconstruction of Religious Thought in Islam*, p. 133-134.

45 Durant, *The Story of Civilization*, Vol. 4, p. 238.

46 Ibid. p. 239.

47 Durant, p. 245.

48 Ibid. p. 289.

49 Ibid. p. 288.

50 Ali, Ameer. *The Spirit Islam*, p. 424

51 Draper, John, *The Intellectual Development of Europe*, Vol. II, p.48.

52 *Encyclopaedia Britannica*, (15th ed.), Micropaedia, "Nazzam, Ibrahim," Vol. VII. p. 233.

53 Hakim, Dr Khalifa Abdul. *The Metaphysics of Rumi*, p. 34.

54 Khaldun, Ibn. *The Muqaddimah*, trans. by Franz Rosenthal, Vol. 1, p. 173.

55 Gould, Stephen Jay. *Ever Since Darwin*. p. 44-45.

56 Darwin, Charles. *The Origin of Species*. p. 49, 74.

57 Biruni, Al. (1000 A.D) *Fi Tahqiq Ma Li'l-Hind (Alberuni's India)*, Translated by Dr.Edward C. Sachu, p. 400.

58 Ibid. p. 400.

59 Darwin, Charles. *The Origin of Species*. p. 75, 86, 88

60 Biruni, Al. *The Athar-Ul-Bakiya (The Chronology ofAncient Nations)*, Translated by Dr.C. Edward Sachu, p. 92-93.

61 Ibid. p. 295.

62 Hakim, Dr Khalifa Abdul. *The Metaphysics of Rumi*, Ibid. p. 39.

63 Ibid. p. 37-38.

64   Gould, Stephen Jay. *Wonderful Life*. New York: W.W. Norton & Company, Inc. 1989, p. 320.

65   Stanfield, William D. *The Science of Evolution*, New York: The Macmillan Company, 1977. p. 26.

66   Ansari, Muhammad Fazl-ur-Rahman. *The Qur'anicFoundations and Structure of Muslim Society*, Vol. II, p. 45.

67   Ali, Yusuf. *The Qur'an, Translation and Commentary*, Verse 29:20.

68   Al-Suhrawaardy, Allama Sir Abdullah. *The Sayings of Muhammad*, p. 94.

NOTES TO CHAPTER 7

1   Khatib, Dr. M. M. *The Bounteous Koran*, Verses 3:190-191.

2   Ansari, Dr.Muhammad Fazl-ur-Rahman. *The Qur'anic Foundations and Structure of Muslim Society*, Vol. II. p. 45.

3   Ali, Yusuf A. *The Qur'an, Translation and Commentary*, Verse 4: 82.

4   Al-Halveti, Sheikh Tosum Bay Bayrack Al-Jerrahi. *The Most Beautiful Names*, p. vii-viii.

5   Asad, Muhammad. *The Message of the Qur'an*, Verses 96: 1-2.

6   Azad, Mawlana Abdul Kalam. *Tarjuman al-Qur'an*, Trans. Syed Abdul Latif, Vol. I. p. 19.

7   Ali, Maulana Muhammad. *The Religion of Islam*, p. 135.

8   Lane, Edward William. *Arabic-English Lexicon*, part 1. p. 7

9   Ibid. Part 2, p. 799.

10   Ibid. Part 2, p. 799.

11   Ibid. Part 2, p. 800.

12   Ibid. Part 2, p. 799-800

13   Ali, Yusuf A. *The Qur'an*, Verses 59:24.

14   Lane. *An Arabic-English Lexicon*. part 1, p. 178.

15   Ali, Yusuf. *The Qur'an*. p.1529

16   Ibid.

17   Ibid.

18   Lane, Edward William. *An Arabic-English Lexicon*, part 4, p. 1745.

19   Ali, Yusuf. *The Qur'an*, Verse 10-4.

20   Ibid. p. 484

21   Ibid. Verse 32:7.

22   Ali, Ahmed. *Al-Qur'an*, Verse 76:2.

23 Bucaille, Maurice. *What is the Origin of Man,* p. 173.

24 Lane, Edward William. *Arabic-English Lexicon,* part 6, p. 2350.

25 Khatib, Dr. M. M. *The Bounteous Koran,* Verse 64:3.

26 Asad, Muhammad. *The Message of the Qur'an,* Verse 71:14.

27 Gould, Stephen Jay. *Ever Since Darwin,* p. 61.

28 Lane, Edward William. *Arabic-English Lexicon.* part 3, p. 1142.

29 Arberry, A. J. *The Koran Interpreted,* p.315.

30 Hakim, Khalifa Abdul. *The Metaphysics of Rumi,* p.36.

31 Shakir, M. H. *Holy Qur'an Translation,* Verses19:9.

32 Ali, Yusuf A. *The Qur'an,* Verse 21:30.

33 Ibid. Verse 23:12.

34 Ibid. Verse 22:5.

35 Lane, Edward William. *An Arabic-English Lexicon,* part 4, p. 1397.

36 Lane, Edward William. *An Arabic-English Lexicon,* part 1, p. 301.

37 Ibid. Part 8. p. 2753

38 Ali, Yusuf A. *The Qur'an,* Verse 24:45.

39 Lane, Edward William. *An Arabic-English Lexicon.* part 3. p. 842.

40 Khaldun, Ibn. *The Muqaddimah.* Vol. 2. p. 424.

41 Pickthall, Mohammed Marmaduke. *Holy Qur'an,* Verse 2: 21.

42 Ibid. Verse 84:19.

43 Ibid. Verse 56:61.

44 Elissi, Muhammad 'Abdul Haleem. *Holy Qur'an,* Verse 27:82.

45 Ali. Yusuf. *The Qur'an,* Verse 28:68.

46 Asad. *The Message of the Qur'an,* Verse 87:1-3.

47 Al-Biruni, *Athar-ul-Bakiya (Chronology of Ancient Nations),* p. 98

48 Al-Jahiz (AD 776-869). *Chance or Creation,* Translated by M. A. S. Abdel Haleem, pp. 113-114.

49 Ibn Khaldun. *The Muquddimah,* Trans. Franz Rosenthal, Vol. 1. p. 195.

50 Russell, Bertrand. *Religion and Science,* p. 80.

51 Ali, Yusuf. *The Qur'an,* Verse 36:82.

52 Sagan, Carl. *Cosmic Connections,* p. 246.

53 Al-Biruni. *Kitab Tahdid Nihayat al-Amakin LitashiahMasafat al-Masakin (The Determination of the Coordinates of Positions for the Correction of Distances between Cities),* Trans by Jamil Ali, p. 15-16.

54 Davies, Paul. *God and the New Physics,* p. 102.

55 Ibid. p. 102.

56 Ibn Khaldun. *The Muqaddimah*, Vol. 1, p. 173.

57 Ali, Yusuf. *The Qur'an*, Verse 15:28-29

58 Shakir, M. H. *Holy Qur'an*, Verse 17:85.

59 Ali, Yusuf. *Holy Qur'an*, Verse 6:103.

60 Milner, Richard. *The Encyclopedia of Evolution.*

NOTES TO CHAPTER 8

1   Khatib, Dr. M. M. *The Bounteous Koran*, Verses 82:6-7.

2   Pickthal, Mohammed Marmaduke. *The Qur'an*, Verse 7:11.

3   Lane, Edward William. *An Arabic-English Lexicon*, Part 5, p. 1973.

4   Ibid. Part 8, p. 1745.

5   Ibid. Part 2, p. 792

6   Kassis, Hanna E. *A Concordance of the Qur'an*, p. 687.

7   Arberry. A. J. *The Koran*, verse 6:133.

8   Lane, Edward William. *An Arabic-English Lexicon*, Part 8, p. 2791.

9   Azad, Mawlana Abul Kalam. *The Tarjuma al-Qur'an.* Hyderbad, Verses 3:33-34.

10 Draper, John William. *History of the IntellectualDevelopment of Europe,* Vol. 1, p. 407.

11 Asad, Muhammad. *The Message of the Qur'an,* Verse 4:1.

12 Maududi, Abu A'la S. *The Meaning of the Qur'an*, Vol. II, p. 94

13 Lane, Edward William. *An Arabic-English Lexicon*, Part 8, p. 2827.

14 Asad, Muhammad. *The Message of the Qur'an*, p. 100.

15 Turpin R, Lejeune J, Lafourcade J, Chigot PL, Salmon C. (1961) *"Presomption de monozygotisme en depit d'un dimorphisme sexuel: sujet masculin XY et sujet haplo X,"* 252:2945-6.

16 Ibn Khaldun. *The Muqaddimah.* Trans.Franz Rosenthal, Vol. I, p. 173.

17 Erwin, Douglas H., and James W. Valantine. "Hopeful Monsters, Transposons and Metazoan Radiation." *Proceedings of National Academy of Science.* Vol.81, p. 5482.

18 Ibn Khaldun. *The Muqaddimah*, Vol. II, p. 444-446.

19 Ibid. Vol. II, p. 423.

20 Ali, Yusuf A. *The Holy Qur'an*, Verse 49:13

21 Ibid. 2:213.

22 Cavalli-Sforza, Luigi Luca. "Genes, Peoples and Languages," *Scientific American,* 265.5: pp. 104-106.

23 Ibid.

24 Shreeve, James. "The Argument Over a Woman," *Discover,* p. 59.

25 Gobons, Ann "Human Evolution:Y Chromosome Shows That Adam Was an African." *Science.* 278.10: pp. 804-805.

26 Strauss, Evelyn. "Can Mitrochondrial Clocks Keep Time?" *Science.* 5 March, 1999: 1438.

27 Cavalli-Sforza, Luigi Luca. "Genes, Peoples and Languages." *Scientific American,* 265.5 (1991): 104-106.

28 Ibid.

29 Allman, William F. "The Origin of Modern Humans. WHO WE WERE." *U.S.News & World Report.* (1991): Vol. 111. No. 12

30 Ali, Yusuf A. *The Holy Qur'an,* Verse 30:22.

NOTES TO CHAPTER 9

1 Niazi, Maulana Kauser. *Creation of Man,* p. 88.

2 Asad, Muhammad. *The Message of the Qur'an,* p. 9.

3 Maududi, Abu A'la S. *The Meaning of the Qur'an,* Verse 32:17.

4 Ibid. Vol. I, p. 51.

5 Ali, Yusuf A. *The Holy Qur'an,* Verses 2:31 and 34.

6 Ibid. Verse 7:13.

7 Maududi, Abu A'la S. *The Meaning of the Qur'an,* Verses 7:14-15.

8 Asad, Muhammad. *The Message of the Qur'an,* Verse 7:20.

9 Ali, Yusuf A. *The Holy Qur'an,* Verse 20:120.

10 Maududi. *The Meaning of the Qur'an,* Verses 37: 7-10.

11 Ibid. Vol. 12, p.65.

12 Ali, Yusuf A. *The Holy Qur'an,* Verse 44:55-56.

13 Arberry, A. J. *The Koran,* Verse 19:62.

14 Ibid. Verses 56:25-6; 78:35; 88:11.

15 Khatib, Dr. M. M. *The Bounteous Koran,* Verse 7:12.

16 Ali, Yusuf A. *The Holy Qur'an,* Verse 38:49.

17 Ibid. Verses 20:118-119.

18 Ibid. Verse 2:30.

19   Ibid. Verse 6:2.

20   Lane, Edward William. *An Arabic-English Lexicon*, Part 1, p. 37.

21   Ansari, Dr Muhammad Dazl-Ur-Rahman. *The Qur'anic Foundations and Structure of Muslim Society*, Vol. II. p. 45.

22   Ali, Yusuf A. *The Holy Qur'an*, Verse 2:38.

23   Asad, Muhammd. *The Message of the Qur'an*, p.205.

NOTES TO CHAPTER 10

1   Ali, Ahmed. *Al-Qur'an*, Verse 67:2.

2   Yusuf Ali A. *The Holy Qur'an*, Verse 2:214.

3   Ibid. Verse 2:286.

4   Rippin, Andrew and Knappert, Jan. *Textual Source for the Study of Islam*, pp. 116-121.

5   Asad, Muhammad. *The Message of the Qur'an*, p.930.

NOTES TO CHAPTER 11

1   Gribbin, John. *In Search of Schrodinger's Cat.* p. 5.

2   Davis, Paul. *God and the New Physics*, pp. 101-102.

3   Russell, Robert John. "Special Providence and Genetic Mutation: A New Defense of Theistic Evolution," *Evolutionary and Molecular Biology*, p. 202.

4   Gribbin, John. *In Search of Schrodinger's Cat.* p. 61.

5   Miller, Kenneth R. *Finding Darwin's God*, p. 209.

6   Nasr, Seyyed Hossein. *Islamic Spirituality Foundations*, p. xxi.

7   Barbour, Ian G. *Religion And Science*, p.187.

8   Russell, Robert J. "Theistic Evolution: Does God really Act in Nature?" *Center for Theology and the Natural Science Bulletin*, 1995, 15.1.

9   Miller, Kenneth R. *Finding Darwin's God*, p. 207.

NOTES TO CHAPTER 12

1   Ali, Ahmed. *Al-Qur'an*, Verse 17: 44.

2   Khatib, Dr. M.M. *The Bounteous Koran*, Verse 41:11.

3   Ibid, Verse 13:13.

4   Khatib, Dr. M. M. *The Bounteous Koran*, Verse 21:69.

5   Khatib, Dr. M. M. *The Bounteous Koran*, Verse 38:36.

6   Hakim, Dr Khalifa Abdul. *The Metaphysics of Rumi*, p. 35.

7   Khatib, Dr. M. M. *The Bounteous Koran*, Verse 33:72

8   Ali. Sayed Anwar. *Qur'an The Fundamental Law of Human Life*, Vol. 11, pp. 362-364.

9   Ibid.

10   Khatib, Dr. M. M. *The Bounteous Koran*, Verse 36:65.

11   Pickthall, Mohammed Marmaduke. *Holy Qur'an*, Verse 41:20-21.

12   Morris, Richard. *The Nature of Reality*, p. 15.

13   Haught, John F. *God After Darwin*, pp. 176-180.

14   Pickthall, Mohammed Marmaduke: *Holy Qur'an*, Verse 8:51.

15   Yusuf Ali A. *The Holy Qur'an*, Verse 13:11.

16   Lovelock, James, E. Gaia, *A New Look at the Earth*.

17   Yusuf Ali A. *The Holy Qur'an*, Verse 41:12.

18   Khatib, Dr. M. M. *The Bounteous Koran*, Verse 55:33.

19   Ibid, Verse 64:11.

20   Ibid, Verses 2:155-157.

21   Ibid, Verse 2:30.

22   Barbour, Ian G. *Religion And Science*, p. 314.

NOTES TO SUMMARY

1   Khatib, Dr. M. M. *The Bounteous Koran*, Verse 17:21.

2   Arberry, Arthur J. *The Koran Interpreted*, Verse 56:25.

3   Barbour, Ian G. *Religion and Science*, p. 296.

4   John F. Haught, *God After Darwin*, p.103

5   Ibid. p. 99.

6   Khatib, Dr. M. M. *The Bounteous Koran*, Verse 2:153.

7   Ibid. Verse 2:156.

8   Ibid. Verse 2:157.

BIBLIOGRAPHY

A

Abelson, Philip H. "Chemical Events on the Primitive Earth." *Proceedings of National Academy of Science*, Vol. 55 (1966), p. 1365-1372.

Adler, Mortimer J. *Aristotle for Everybody*. New York: Macmillan Publishing Co. Inc, 1978.

Al-Biruni. (d. 1048 AD.) *The Athar-Ul-Bakiya (The Chronology of Ancient Nations)*. Translated by Dr. C. Edward Sachu. London: W. H. Allen, 1879.

_____. *Fi Tahqiq Ma Li'l-Hind (Alberuni's India)*. Translated by Dr. Edward C. Sachu. London: Kegan Paul, Trench, Trubner & Co. Ltd., 1914.

_____. *Kitab al-Jamahir fi ma'rifat al-jawahir,* (Arabic). Hyderabad: Dairatu'l Ma'arif'l Osmania, 1934.

_____. *Kitab Tahdid al-Amakin Listashiah Masafat al-Masakin (The Determination of the Coordinates of Positions for Correction of Distances between Cities)*. Translated by Jamil Ali. Beirut: The American University of Beirut, 1967. al-Faruqi, I. R. *The Cultural Atlas of Islam*. New York: Macmillan Publishing Company, 1986.

Al-Ghazzali, Imam Abu-Hamid. (1058-1111 AD.) *Ihya Ulum-Id-Din*. Translated by Al-Haj Maulana Fazal-Ul-Karim. Lahore. Pakistan: Book Lovers Bureau, 1971.

Ali, Ahmed. *Al-Qur'an*. Princeton: Princeton University Press, 1990.

Ali, Ameer. *The Spirit of Islam*. Lahore, Pakistan: Pakistan Publishing House, 1981.

Ali, Maulana Muhammad. *The Religion of Islam*. Lahore, Pakistan: Ahmadiyyah Anjuman Ish'at Islam, 1950.

Ali, Sayed Anwar. *Qur'an, The Fundamental Law of Human Life*. Karachi, Pakistan: Hamdard Foundation Press, 1987.

Ali, Yusuf. *The Holy Qur'an*. Indianapolis: American Trust Publications, 1977.

Al-Halveti, Sheikh Tosum Bay Bayrack Al-Jerrahi. *The Most Beautiful Names*. Vermont: Threshold Books, 1985.

Al-Jahiz. (AD 776-869.) *Chance or Creation*. Translated by M. A. S. Abdel Haleem. Berkshire, U.K: Garnet Publishing Limited, 1995.

Allman, William F. "*The Origin of Modern Humans*. WHO WE WERE." U.S. News & World Report. Vol. 111 (1991), No. 12.

Al-Suhrawaardy, Allama Sir Abdullah. *The Sayings of Muhammad*. New York: Carol Publishing Group, 1990.

Ansari, Muhammad Fazl-ur-Rahman. *The Qur'anicFoundations and Structure of Muslim Society*. Karachi, Pakistan: The World Foundation of Islamic Mission, 1977.

Arberry, Arthur J. *The Koran interpreted*. New York: Collier Books, 1955.

Asad, Muhammad. *The Message of the Qur'an*. Gibraltar: Dar Al-Andalus, 1980.

Ashley, Montagu, ed. *Science and Creationism*. Oxford, New York, Toronto, Melbourne: Oxford University Press, 1984.

Asimov, Isaac. *Beginning: The Story of Origins—of Mankind, Life, the Earth, the Universe*. New York: Walker & Co., 1987.

Azad, Mawlana Abdul Kalam. *Tarjuman al-Qur'an*. Translated by Syed Abdul Latif. Hyderabad, India: Dr. Syed Abdul Latif's Trust for Qur'anic and Other Cultural Studies, 1981.

B

Barbour, Ian G. *Religion and Science*. New York: HarperCollins, 1997.

Barrow, John D., and Joseph Silk. *The Left Hand of Creation*. New York: Basic Books Inc, 1983.

Blackmore, Vernon, and Andrew Page. *Evolution the Great Debate*. Illinois: Lion Publishing Company, 1989.

Briffault, Robert. *The Making of Mankind*. Lahore, Pakistan: Islamic Book Foundation, 1980.

Bronowski, Jacob. *The Ascent of Man*. London, U.K: The British Broadcasting Corporation, 1981.

Bucaille, Maurice. *What is the Origin of Man?* Paris, France: Seghers, 1983.

Bukhari, Muhammad ibn Ismail al-Jufi. (d. 897). *Sahih Al-Bukhari*. Translated by Muhammd Muhsin Khan. Chicago: Khazi Publications, 1977.

C

Cavalli-Sforza, Luigi Luca. "Genes, Peoples and Languages." *Scientific American*. 265.5 (1991): 104-106.

Curtis, Helena and Sue N. Barnes. Biology, (5th ed.). New York: Worth Publishers, 1989.

D

Daniel Dennett. *Darwin's Dangerous Idea: Evolution and the meaning of life*. New York: Simon & Schuster, 1995.

Darwin, Charles. *The Origin of Species*. New York, Scarborough: New American Library, 1958.

Davies, Paul. *God and the New Physics*. Simon and Schuster, Inc., 1983.

Dawood, N. J. *The Koran*. New York: Penguin Books, 1990.

Dimroth, Erich, and Michael M Kimberley. "Pre-Cambrian atmospheric oxygen: Evidence in the sedimentary distributions of Carbon, Sulfur, Uranium, and Iron." *Canadian Journal of Earth Sciences*, XIII (September 1976): 1161-1185.

Dods, D.D., Marcus, and others. "Book of Genesis" in *An Exposition of the Bible*. Hartford: The S. S. Scranton Co, Vol. 1. 1903.

Dodson, Edward O., and Peter Dodson. *Evolution: Process Product*. Boston: Prindle, Weber & Schmidt, 1985.

Draper, John William. *The Conflict between Religion and Science.* New York: D. Appleton and Company, 1875.

Draper, William John. *The Intellectual Development of Europe.* New York: London: Harper & Brother, 1876.

Durant, Will. *The Story of Civilization.* Vol. 4. New York: Simon and Schuster, Inc., 1950.

E

Eaton, C. G. *Islam and the Destiny of Man.* New York: Macmillan, 1985.

Edey, Maitland A., and Donald C. Johanson. *Blue Print.* Boston, Toronto, London: Little, Brown and Company, 1989.

Elissi, Muhammad 'Abdul Haleem. *Holy Qur'an.* Hyderbad, India: The Barney Academy, 1981.

Erwin, Douglas H., and James W. Valentine. "Hopeful Monsters, Transpons, and Metazoan Radiation." *Proceedings of National Academy of Science.* Vol.81: 5482-5483, 1984.

Encyclopaedia Britannica-Micropaedia. 15th ed. 1981.

Encyclopaedia Britannica-Macropaedia. 15th ed. 1981.

Eve, Raymond A., and Francis Harrold. *The Creationist Movement in Modern America.* Boston: Twayne Publishers, 1991.

F

Ferris, Timothy (Narrator). "*The Creation of the Universe.*" (PBS Home Video. Beverly Hills, 1990.

Futuyma, Douglas J. "World Without Design." *Natural History.* (March, 1987): 34-36.

————. "The Theory of Evolution." *Encyclopaedia Britannica.* 15th ed., pp 981-1009, 1988.

G

Gobons, Ann. "Human Evolution: Y Chromosome Shows That Adam Was an African." *Science.* 278.10 (1997): 804-805.

Godfrey, Laurie R., ed. *Scientists Confront Creationism.* New York, London: W. W. Norton & Company, 1983.

Gould, Stephen Jay. *Time's Arrow Time's Cycle*. Cambridge, Massachusetts, London, U.K: Harvard University Press, 1987.

———. "A quahog is a quahog." *Natural History*, (August-September 1979): 18-26.

Gould, Stephen Jay, and Nile Eldredge. "Puctuated Equilibrium Comes off Age." *Nature* 336(1993): 223-227.

Gould, Stephen Jay. *Eight Little Piggies*. New York, London: W. W. Norton & Company, 1993.

———. *Ever Since Darwin*. New York, London: W. W. Norton, 1977.

———. "Knight Takes Bishop?" *Natural History*, (May 1986): 18-33.

———. *Wonderful Life*. New York: W. W. Norton & Company, 1989.

Guillanme, A. *The Life of Muhammad*. Oxford: Oxford University Press, 1955.

H

Hakim, Dr Khalifa Abdul. *The Metaphysics of Rumi*. Lahore, Pakistan: Institute of Islamic Culture, 1977.

Hardy, Edward R. "Origin and Evolution of Universe," *Encyclopaedia Britannica—Macropaedia*, 15th ed., Vol. 18: 1007-1011.

Haught, John F. *God After Darwin*. Boulder: Westview Press, 1999.

Hawking, Stephen W. *A Brief History of Time*. Toronto, London, New York, Sydney: Bantam Books, 1988.

Henderson-Sellers A., A. Benlow, and A. J. Meadows. "The Early Atmospheres of Terrestrial Planets." *Quarterly Journal of Royal Astronomical Society*. Vo.21 (1980): 74-81.

Hitti, Philip K. *History of Arabs*. New York: St. Martin's Press, 1970.

Hussani, S. A. Q. *The Pantheistic Monism of Ibn Arabi*. Lahore, Pakistan: SH. Muhammad Ashraf, 1979.

I

Ibn al-Hajjaj, Abul Husain Muslim (d. 875). *Sahih Muslim*. Book 18, No. 4261. Translated by Abdul Hamid Siddiqui. New Delhi: Kitab Bhavan, 2000. (Available on the Web: *www.usc.edu/dept/MSA/fundamentals/hadithsunnah/muslim/*)

Ibn-Khaldun, Abdul-Rahman (d. 1406). *Muqaddimah*. Translated by Franz Rosenthal. Princeton: Princeton University Press, 1980.
_____. *The Muqaddimah* (Abridged edition). Translated by Franz Rosenthal. Princeton: Princeton University Press.
Ikhwan al-Safa. *Rasa'il Ikhwan al-Safa wa-Khull an al-Wafa ("Epistles of Brothrens of Purity")*, (Arabic). Beirut, Lebanon: Dar-ul-Beirut Publishers, 1983.
Iqbal, Sir Mohammad. *The Reconstruction of Religious Thought in Islam*. Lahore, Pakistan: Ashraf Printing Press, 1982.
Irving, T. B. *The Qur'an*. Brattleboro: Amana Books, 1985.

J

Jastrow, Robert. *God and the Astronomers*. New York: Warner Books, 1908.

K

Kaufman III, William J. *Black Holes and Warped Spacetime*. New York: W. H. Freedman and Company, 1979.
Kassis, Hanna E. *A Concordance of the Qur'an*. Berkeley, Los Angeles, and London: University of California Press, 1983.
Khathib, M. M. *The Bounteous Koran*. London: Macmillan Press, 1984.

L

Lane, Edward William. *Arabic-English Lexicon*. Beirut, Lebanon: Libraire Du Liban, 1980.
Lovelock, James E. *Gaia, a New Look at the Earth*. London: Oxford University Press, 1974.
Lucas, J. R. *Historical Journal*. XXII (1979): 313-30.

M

March, Robert H. *Physics for Poets*. Chicago: Contemporary Books, 1978.
Marvin, Ursula B. "Continental Drift." *EncyclopaediaBritannica*, Vol. 5:108-115.

Morris, Richard. *Time's Arrows*. New York: Simon and Schuster, Inc., 1986.

Morris, Richard. *The Nature of Reality*. New York: Noonday Press, 1988.

Maududi, Abul A'la S. *The Meaning of the Qur'an*. Lahore, Pakistan: Islamic Publications (Pvt.) Ltd., 1984.

Mayr, Ernst. *One Long Argument: Charles Darwin and the Genesis of Modern Evolutionary Thought*. Cambridge: Harvard University Press, 1991.

Mayr, Ernst, William Provine, eds. *The Evolutionary Synthesis*. "Prologue: Some Thought on the History of the Evolutionary Synthesis." Cambridge, Massachussetts, and London, England: Harvard University Press, 1981.

Miller, Stanley L., and Robert G. Orgel. *The Origin of Life on the Earth*. Englewood Cliffs: Prentice-Hall, Inc., 1969.

Miller, Kenneth. *Finding Darwin's God*. New York: HarperCollins Publishers Inc., 1999.

Miller, Richard. *The Encyclopedia of Evolution*. New York: Henry Holt and Company, Inc., 1990.

Moyers, Bill. *A World of Ideas*. Edited by Betty Sue Flowers. New York: Doubleday, 1989.

N

Nasr, Seyyed Hussein. *Introduction, in Islamic Spirituality Foundations*. New York: The Crossroad Publishing Company, 1991.

Nasr, Seyyed Hossein. "Islam and the Environmental Crisis." *MAAS Journal of Islamic Science*. Vol. 6. No.2 (1990): 31-51.

Niazi, Maulana Kauser. *Creation of Man*. Lahore, Pakistan: S. H. Muhammad Ashraf, 1974.

Nicholson, Reynold A. *Rumi*. San Francisco: Divani Shamsi Tabriz, 1973.

O

Osborn, H. F. *From Greeks to Darwin*. London and New York: The Macmillan Company, 1922.

P

Pickthall, Mohammed Marmaduke. *Holy Qur'an*. Hyderbad, India: H. E. H. Mir Osman Ali Khan, 1930.

Plott, John C. *The Global History of Philosophy*. Delhi, India: Motilal Banaridass, 1984.

R

Raup, David M. "Conflict between Darwin and Paleontology." *Field Museum of Natural History Bulletin*, (January 1979): 22-29.

Richard Dawkins. *The Blind Watchmaker*. New York: W.W. Norton, 1986.

_____. *River out of Eden*. New York: Basic Books, 1995.

Richard Lewontin. A Review of Carl Sagan's book *The Demon-Haunted World: Science as a Cradle in the Dark*, New York Review of Books. (January 9, 1997): 28-32.

Rippin, Andrew and Jan Knappert. *Textual Source for the Study of Islam*. Chicago: The University of Chicago Press, 1990.

Russell, Bertrand. *Religion and Science*. London, New York: Oxford University Press, 1961.

Russell, Robert J. "Theistic Evolution: Does God really Act in Nature?" *Center for Theology and the Natural Science Bulletin*, 15.1, 1995.

Russell, Robert J; William R Stroeger, and Francisco J.Ayala. *Evolution and Molecular Biology*. "Special Providence and Genetic Mutation: A New Defense of Theistic Evolution," by Robert J. Russell. Vatican City State, Berkeley: Vatican Observatory Foundation and The Center for Theology and Natural Sciences, 1998.

S

Sagan, Carl. *Cosmos*. New York: Random House, 1980.

_____. *Cosmic Connections*. New York: Anchor Books, 1973.

_____. "Life." *Encyclopaedia Britannica* 15th ed., Vol. 10:893-911.

Salahi, Adil M., and Ashur A Shamis. *In the Shade of the Qur'an*, Vol. 30. London: M W H London Publisher, 1979.

Schroeder, Gerald L. *The Science of God:The Convergence of Scientific Biblical Wisdom*. New York: The Free Press, a Division of Simon & Scuster Adult Publishing Group, Copyright © 1977 by Gerald L. Schroeder. Reprinted with permission of The Free Press.

————. *Genesis and Big Bang*. New York, Toronto, London, Sydney, Aukland: Bantam Books, 1990.

Shakir, M. H. *Holy Qur'an Translation*. New York: Tahrike Tarsile Qur'an, Inc., 1985.

Shapiro, Robert. "Probing of the Origin of Life." *Encyclopaedia Britannica: Yearbook of Science and Future*. Chicago, 1984.

Shadewald, Robert. *National Center for Science Education Reports*. (March-April 1990) Vol. 10: 21.

Shermer, Michael. "25 Creationists' Arguments & 25 Evolutionists' Answers." *Skeptic*. Vol. 2, No. 2.

Smart, John Jamieson Carswell. "Time," *Encyclopaedia Britannica*. 1981, 15th ed. Vol.18: 410-421.

Stanfield, William D. *The Science of Evolution*. New York: The Macmillan Company, 1977.

Stanley, Steven M. *Macroevolution Pattern and Process*. San Francisco: W. H. Freeman and Company, 1979.

Shreeve, James. "The Argument over a Woman." *Discover*. 2.8 (1990): 52-59.

Strauss, Evelyn. "Can Mitochondrial Clocks Keep Time?" *Science*. (5 March, 1999) Vol. 283: 1436-1438.

T

Trefil, James. *Reading the Mind of Go: In Search of the Principle of Universality*. New York, London, Toronto, Aukland: Doubleday, 1989.

Tufail, Abu Bakr Muhammad. *Hai bin Yaqzan* (The Journey of the Soul). Translated by Riad Kocache. London:U.K: The Octagon Press, 1982.

Turpin R, Lejeune J, Lafourcade J, Chigot PL, Salmon C. (1961) "Presomption de monozygotisme en dépit d'un dimorphisme sexuel: sujet masculin XY et sujet neuter haplo X." *C R Acad Sci[D]* (Paris), Vol. 252:2945-6.

W

Whinfield, E. H. *Teachings of Rumi*. London: The Octagon Press, 1984.

Watch Tower Bible &Tract Society of Pennsylvania (New York). *Did Man Get Here by Evolution or Creation?* 1967.

Y

Yahya, Harun. *The Evolution Deceit*. London, UK: Ta-ha Publishers Ltd., 1999.

GLOSSARY

Adaptation: Physical structure, bodily function (physiology), or behavioral characteristic that strengthens the ability of the organism to cope with its prevailing environment.

Adaptive radiation: Tendency of successful species (or higher groups) to spread into all available ecological niches.

Alga (plural: algae): Any of a broad group of simple, mostly aquatic plants.

Al-Akhirah: The Arabic word for the hereafter. According to Muslim belief our present universe will be completely disposed of and a new universe will be created into which mankind will be resurrected.

Allah: The Arabic term for God, the Supreme Creator of all that exists. There is no plural for the word Allah in Arabic. He has no partners, sons or daughters. He is neither male nor female. He is self-sufficient and no deities precede or exceed him.

Allele: An alternative form of genes occupying the same position (locus) in a homologous chromosome.

Allopatric: Pertains to populations that are geographically separated from one another.

Allopatric speciation: development of new species from single parent species that live in separate territories (allopatric territories).

Amino acid: The basic structural unit of protein.

Amphibian: Any intermediate group of animals between fish and reptiles.

Anagenesis: Transformation of whole populations in the evolution of life.

Analogy (adj.: analogous): Similarity in function but difference in structure and origin (e.g., wings of insects and birds).

Annelid: A group of organisms with tiny rings and many body segments fused together in a linear arrangement (earthworms, leeches). They have closed circulatory systems.

Annual ring: Annual variations in the growth rate of trees which can be seen in the series of concentric rings of cross-sections of trees.

Apogee: The most distant point from the center of the earth in the orbit of the moon or of an earth-orbiting artificial satellite.

Arthropod: Segmented animals with paired jointed legs and stiff external skeletons (crabs, lobsters, insects, spiders).

Atom: The smallest part of a chemical element that can take part in any chemical reaction and still retain its identity. All atoms consist of a nucleus and one or more orbiting electrons.

Australopithecus: A genus of fossil hominids found in southern and eastern Africa dating from the Late Pilocene to the end of the Early Pleistocene (at least 2,500,000 years). It is often considered a possible ancestral genus to modern man.

Al-Baari: One of the epithets of God in the Qur'an. An Arabic word derived from the verb *Baara,* which means "evolving from a previously created matter or state." The noun *Al-Baari* means "evolver."

Bacterium: Microscopic unicellular organisms with simple DNA molecules not contained in a nuclear membrane.

Big Bang: The theory that the emergence of our universe was the result of the expansion of matter, space and time, following an explosion of the "cosmic egg" or the initial state of enormous density and pressure at the beginning of the universe.

Big Crunch: Following the Big Bang the universe continues to expand. At some time in the future the universe will start collapsing into a singularity which will lead to its end. This process of corrosion is called the Big Crunch.

Black hole: When a large star implodes due to its own gravitational pull, even the light cannot escape from it. As a result, a black hole is created out of which an observer cannot see the events inside that hole.

Bosons: Subatomic particles we often think of as waves, or radiation. The weak boson carries the weak force when the atomic nuclei decay.

Competition: struggle between members of different species for a mutually required and limited resource.

Chromosome: A long strand of coiled DNA in the nucleus of a cell; threaded around protein. Chromosomes are made up of genes, and each species is classified by its unique number of genes.

Class: A taxonomic grouping of organisms belonging to related orders; the major subdivision of a phylum.

Condensation reaction: A chemical process which takes place when two monomers join. A monomer is a small molecule from which a polymer is derived.

Conifer: A cone-bearing plant.

Coordinates: Mathematical ways of delineating positions.

Cosmic string: When the temperature reaches the so-called Planck value of 10^{32} degrees (1 followed by 32 zeros), all matter is dissociated into its most primitive constituents, which are called cosmic strings.

Cro-Magnon man: Prehistoric beings considered by some scholars to be the ancestors of modern mankind.

Cytoplasm: The semifluid substance that makes up the nonnuclear part of a cell.

Deoxyribonucleic acid (DNA): The chemical substance in the chromosome the structural arrangement of which is the basis of inheritable characteristics.

Dizygotic: developed from two fertilized eggs (zygotes), as a fraternal twin.

Doppler effect: The displacement of spectral lines (dark lines in a spectrum) in the radiation received from a source due to its relative motion along the line of sight. An approaching motion is indicated by a blue shift; a receding motion is indicated by a red shift.

Dryopithecus: A genus of extinct apelike animals representative of the dryopithecines (a group of small, generalized apes that contains the ancestors of both the modern apes and humans).

Ecology: Study of the interplay of living organisms and their environment.

Echinoderm: A form of marine life with pentaradial symmetry, a calcareous (calcium containing) skeleton, a water vascular system and tubed feet (starfish, sea cucumber, sea urchins).

Ecological isolation: Presence of genetic differences between populations that prevent interbreeding.

Electromagnetic force: The force that operates between particles with electric charge. This is the second strongest of the four fundamental forces.

Electron: A sub-particle with a negative electric charge that orbits the nucleus of an atom.

Ethology: Study of patterns of behavior of animals in their natural environments, with special attention to adaptive and comparative aspects

Event horizon: The periphery of a black hole.

Evolution: Any changes in the frequency of alleles within a gene pool from one generation to another that results in the divergence of related populations and gives rise to new species and higher forms of life.

Fertilization: Union of male and female gametes to form the single cell (zygote) from which the embryo develops.

Fossil: Preserved remains or traces of an organism that lived in the past.

Fundamentalism: A militantly conservative movement in American Protestantism originating around the beginning of the twentieth century in opposition to modernist tendencies. It emphasizes the literal interpretation and the absolute authenticity and accuracy of the Scriptures, the imminent and physical second coming of Jesus Christ, the virgin birth, physical resurrection, and substitutionary atonement.

Fundamentalist: An adherent of Protestant fundamentalism.

Gene: The fundamental unit of heredity with specific sequence nucleotides carried in the chromosomes.

Gene flow: Exchange of genes between local and immigrant populations through interbreeding.

Gene pool: All of the alleles of all the genes carried by individuals in a population at a particular time.

Genetic distance: The extent to which two populations differ consistently in their alleles.

Genetic drift: Change in the genetic composition of a small population resulting from chance event, or from sampling error such as meiosis and random fertilization.

Genetic equilibrium: Invariability of particular frequencies of allelic members of a gene through successive generations in a population.

Genetic isolate: A self-containing breeding population that does not exchange genes with any other group.

Genotype: The full set of genes of an organism's chromosome, as distinct from his or her outward appearance (phenotype).

Genus (plural: genera): A taxonomic grouping of very similar organisms considered to be closely related species.

Geographic barrier: A geographical feature (such as mountain or river) that prevents gene flow between populations.

Gluon: A subatomic particle of zero mass. Quarks interact through the exchange of gluons that carry the strong nuclear force.

Grand Unified Theory: A theory that aims to unify the basic forces of nature, bringing into one scheme the strong and weak forces, electromagnetism and gravity. At the extraordinarily high temperatures prevailing in the earliest moments of the Big Bang, these forces were indistinguishable from each other.

Gravitation: The force that keeps the entire universe together by mutual attraction.

Graviton: A massless subatomic particle. The graviton carries the force of gravitation.

Hadiths: The reports of what Prophet said and practiced.

Hemoglobin: Iron-containing protein in vertebrate blood that binds and transports respiratory gases (oxygen and carbon dioxide).

Heredity: Genetic transmission of traits from parent to offspring.

Hominid: Member of the family of upright, bipedal primates that includes modern humans and their related ancestors.

Homo erectus: The extinct species of hominids (the family of man), dating from the middle Pleistocene time (2,500,000 to 10,000 years ago).

Homology: Similarity of structures in a series of related organisms because of descent from common ancestors, without regard to function.

Homologous structure: Body parts that are similar in structure owing to common evolutionary and developmental origins.

Homo sapiens: The species to which all modern human beings belong.

Homo sapien sapiens: Modern human.

Iblis: The Arabic proper name for the Devil.

Immunity: The ability of an organism to resist or overcome an infectious agent or its antigens by the presence or production of antibodies.

Inbreeding: The mating of related individuals with similarities in their genetic makeup. This mating occurs in greater or lesser frequency than would be predicted by chance.

In vitro: within a glass; observable in culture or in a test tube.

Isolating mechanisms: A morphological, behavioral, physiological, or geographical barrier that prevents or limits gene exchange between different species.

Jennat-ul-Khuld: The Arabic term for Eternal Paradise; the place of reward and final return for those who are righteous in their worldly thoughts and actions.

Kelvin scale of temperature: A scale of temperature whose units (called kelvin, symbol K) are equal in size to those of the Celsius scale, whose zero is fixed at -273.16°C, often known as absolute zero.

Khalaqa: A verb used in almost all verses of the Qur'an in that refer to the creation of mankind. It means "to bring into existence according to a certain measure or proportion a being, and to make it equal to another thing which is not like any preexisting being."

Linkage: An association of genes of different characteristics in the same chromosome. In the absence of crossing over, linked genes do not assort randomly.

Mammals: The highest class of vertebrate animals that possess hair and suckle their young.

Masnavi: A collection of the spiritual couplets of Jalaluddin Rumi, which has greatly influenced Muslim mystical thoughts and literature.

Meiosis: the cellular process that results in the number of chromosomes in the gamete-producing cell being reduced to one half (the process of the formation of sperm and egg).

Melanin: Dark brown or black pigment of the skin or hair.

Micro-organism: Minute, usually microscopic, living organisms such as bacteria, viruses, and yeast.

Milky Way: The name of a great arc of stars and dust, seen in clear skies from most latitudes, in or near the plane of the galaxy in which the Sun, Earth, and other members of the solar system are located.

Monomer: small molecules from which a polymer is derived.

Mono-zygote: Developed from a single fertilized egg (zygote) as identical twin.

Musawwir: An Arabic epithet of God used in the Qur'an. It means: "Fashioner, of all existing things. The One who created them and gave each of them a special form and specific manner of being whereby the can be distinguished from others." In modern biological vernacular, it could mean "someone who causes speciation."

Mutation: An alteration of the gene, by deletion, addition, or substitution of one or more nucleotide bases of DNA strand, which results in an inheritable change in its traits.

Natural selection: Differential reproduction of offspring (genotype) from one generation to the next.

Neanderthal: A type of prehistoric creature who inhabited much of Europe and the areas surrounding the Mediterranean during the earlier Late Pleistocene Epoch. The Neanderthals were cave dwellers, short but powerfully built.

Niche: The way of life of a species that includes every aspect of its ecological, functional, and behavioral roles.

Nuclear decay or radioactivity: The spontaneous breakdown of an unstable atomic nucleus with the emission of particles and rays.

Nucleic acid: long-chain polymer of nucleotides.

Nucleotide: A basic repeating unit in a nucleic acid consisting of a 5-carbon sugar molecule, a phosphoric acid, and a ring-shaped nitrogenous base.

Nucleus: In atomic physics the nucleus is the central part of an atom, consisting only of protons and neutrons, held together by a strong force. In biology, the nucleus is the spheroid body within the cell that is bound by a double membrane and contains the chromosomes.

Neutron: An uncharged particle in the nucleus.

Neutrino: A subatomic particle with no mass and no electric charge; it enters only into reactions involving the weak nuclear force.

Orbit: The path of one body around another.

Order: A taxonomic grouping or organisms belonging to similar families.

Organic compound: A chemical compound that has one or more carbon atoms.

Oxidation: The removal of electrons from an atom or compound; oxidation in biological systems generally involves removal of hydrogen electrons.

Paleontology: The study of the past through its fossil remains. A practitioner of the science of paleontology is called paleontologist.

Phalanx: A bone of a finger or toe.

Phenotype: outward (observable) manifestation of individual genetic makeup.

Photon: A quantum particle of light and the messenger particle of electromagnetic radiation (a particle of visible light).

Phyletic gradualism: The notion that evolutionary change is slow and steady.

Phylum: The broadest grouping below kingdom, characterized by a structural plan that differs fundamentally from and is difficult to derive from any other.

Polymer: Large molecule made up of many repeating or like sub-units called monomers.

Population: Individuals of a species that form a local breeding community in a particular region.

Proton: The positively charged particles that make up roughly half the particles in the nucleus of most atoms.

Protozoa: Any of the single-celled organisms.

Punctuated equilibrium: The concept that species (and higher groups) remain stable for long periods of time (equilibrium), then change rapidly (punctuation).

Quantum: The indivisible unit in which waves may be emitted or absorbed.

Quark: A charged elementary particle that feels the strong force. Protons and neutrons are each composed of three quarks.

Qur'an: Islam's book of spiritual guidance. Muslims believe that the Qur'an, in its original Arabic vernacular, is the word of God, revealed to the Prophet Muhammed over a period of twenty years. Translations and commentaries of the Qur'an in any language reflect the translators' interpretations and should not be confused with the Holy Book.

Rabb: An Arabic epithet of God used in the Qur'an. Its literal meaning is "someone who fosters a thing in such a manner as to make it attain one condition after another until it reaches its goal of perfection."

Radioactivity: The spontaneous breakdown of the nuclei of isotopes into the nuclei of other atoms, accompanied by the emission of energy in the form of atomic particles. This breakdown or decay process of nucleus is random, for each kind of radioactive nucleus there is a specific time period, called half time, which describes its rate of decay. In a large sample of nuclei, almost exactly half will have decayed in one half life.

Radioactive dating: The radioactive elements provide accurate nuclear clocks; by comparing the relative abundances of a remaining radioactive element and the element it decays to, we can learn how long the process has been going on and hence arrive at the age of the sample.

Rakah: A fixed series of ritual movements and recitation that are repeated in the daily mandatory prayers of Muslims.

Ramapithecus: This is a fossil primate genus dating from the late Miocene or early Pliocene Epoch classified as a member of the family Hominidae, which includes Australopithecus and Homo, the genus to which modern mankind belongs.

Random mating: Pattern of breeding in which an individual of one genetic makeup has an equal probability of mating with any individual of another genetic constitution.

Rinonucleic acid (RNA: Nucleic acid based upon ribose; occurs in three forms: messenger RNA (mRNA), Ribosome RNA (rRNA), and transfer RNA (tRNA).

Red shift: The reddening of light from a star that is moving away from us, due to the Doppler effect.

Reducing atmosphere: Atmosphere without oxygen.

Reduction: The gain of an atom or molecule by an electron in a chemical reaction.

Regulatory gene: A gene that controls the expression of one or more other genes.

Reproductive isolation: Inherent inability of members of different breeding populations (particularly different species) to breed with each other.

Ribosome: A small organelle in the cell. It is the site of protein synthesis.

Scientific method: The organized approach to the study of natural phenomenon. The initial step in the process is observation and it is followed by the development of hypothesis, systematic experimentation to test the hypothesis, and finally the formulation of an explanation for natural phenomenon based on the results of the experiment.

Sedimentary rock: Rock formed of sediments (particles) of chemical or organic origin.

Singularity: Where space-time is reduced to a mathematical point and the density of matter becomes infinite.

Species: A group of similar organisms whose members can breed with one another to produce fertile offspring.

Stratigraphy: The study of layers of rocks, particularly with respect to the relative ages of the fossils they contain.

Strong force: The strongest of the four fundamental forces with the shortest range of all. It holds quarks together within protons and neutrons and holds protons and neutrons together to form atoms.

Subspecies: A geographical race of a species.

Taxon (plural, taxa): A group of organisms such as species or class as defined by the biological classification scheme (taxonomy).

Taxonomy: Study of classification of organisms into appropriate categories (taxa) on the basis of evolutionary relationships among them. The system consists of a hierarchy of categories. The most inclusive categories are the kingdoms. Other main categories, in descending order, are phylum, class, order, family, genus, and species.

Theory: Hypothesis that is supported by extensive observations and experiments.

Time dilation: The time between two events (the sending of light and its reception at the receiver) is greater for an earth-bound observer than for a traveling observer. This is a general result of the theory of relativity. Its effect is that time is actually measured to pass more slowly by the clock of a traveler (astronaut) than by the clock of a stationary person (earth-bound person).

Uniformitarianism: A theory that that finds all features of the present earth can be explained in terms of the accumulated effects of processes that we can see and measure here and now.

Vertebrate: Animal possessing a backbone.

Vestigial organs: Refers to an apparently functionless or imperfectly developed structure or organ that was fully developed and functional in ancestral species.

Weak force: One of the four fundamental forces experienced by elementary particles except the force-carrying particles. It is responsible for radioactivity.

Xylem: Complex vascular tissue that conducts water and minerals from the roots to other parts of plants.

Zygote: Diploid cell resulting from the union of two haploid gametes (fertilized eggs).

Suggested Books for Further Reading

SCIENCE AND RELIGION

"THE ORIGIN OF SPECIES" by Charles Darwin. Publisher: NAL Penguin Inc., NewYork.

"EVOLUTION PROCESS AND PRODUCT" by Edward O. Dodson and Peter Dodson, Publisher: Prindle, Weber & Schmidt, Boston.

"EVER SINCE DARWIN" by Stephen Jay Gould, Publisher: W/ .W.Norton & Company, New York.

"GOD AFTER DARWIN" by John F. Haught. Publisher: Westview Press, Boulder, Colorado.

"RELIGION AND SCIENCE" by Barbour, Ian G. HarperCollins, New York, NY.

"FINDING DARWIN'S GOD" by Kenneth R. Miller. Publisher: Cliff Street Books, New York, NY.

"ONE LONG ARGUMENT" by Earnest Mayr, Publisher: Harvard University Press, Cambridge, Massachusetts.

"ORIGINS" by Robert Shapiro, Publisher: Bantam Books, New York.

"COSMOS, BIOS, THEOS" edited byHary Margenau and Roy Abraham Varghese, Publisher: Open Court, La Salle, Illinois.

"THE CREATIONISM MOVEMENT IN MODERN AMERICA" by Raymond A. Eve and Francis B. Harrord, Publisher: Tawaye Publisher, Boston.

"SCIENTISTS CONFRONT CREATIONISM" edited by Laurie R. Godfrey, Publisher: W.W.Norton & Co, New York.

"SCIENCE AND CREATIONISM" edited by Ashley Montagu, Published by Oxford University Press, New York.

COSMOLOGY

"COSMOS" by Carl Sagan, Publisher: Random House, New York.

"READING THE MIND OF GOD" by James Trefil, Publisher: Anchor Books, New York.

"THE LEFT HAND OF CREATION" by John D.Barrow and Joseph Silk, Publisher: Basic Books, Inc., New York.

"GOD AND THE NEW PHYSICS" by Paul Davies, Publisher: Simon & Schuster, Inc., New York.

"A BRIEF HISTORY OF TIME" by Stephen W. Hawking, Publisher: Bantam Books, New York.

"THE ABC OF RELATIVITY" by Bertrand Russell, Publisher: New American Library, New York.

"ABOUT TIME, Einstein's Unfinished Revolution" by Paul Davies, Publisher: Simon & Schuster, New York.

"THE LAST THREE MINUTES" by Paul Davies, Publisher: Basic Books, Inc., New York.

ISLAM AND SCIENCE

"AN INTRODUCTION TO ISLAMIC COSMOLOGICAL DOCTRINES" by Seyyed Hossein Nasr, Publisher: Thames and Hudson Ltd., Bath, UK.

"ISLAMIC SCIENCE AN ILLUSTRATED STUDY" by Seyyed Hossein Nasr, Publisher: World of Islam Festival Publishing Company, Ltd.

"SCIENCE AND CIVILIZATION IN ISLAM" by Seyyed Hossein Nasr, Publisher: Barnes and Noble Books, New York.

"ISLAM AND ECOLOGY" ed. by Fazlun Khalid with Joanne O'Brien. Publisher. Cassell Publishers Limited, New York

"THE GENIUS OF ARAB CIVILIZATION Source of Renaissance" ed. John R Hayes, published by New York University Presss, New York.

ISLAM AND ITS THEOLOGY

"THE MESSAGE OF THE QUR'AN" by Muhammad Asad, Publisher: Multimedia Vera International, 434 South Vermont Ave, Los Angeles, CA 90020

"THE HOLY QUR'AN, Translation and Commentary" by Yusuf Ali. Publisher: American Trust Publications, Indianapolis, IN. 1977.

"THE SAYINGS OF MUHAMMAD" by Allama Sir Abdullah Al-Mamum Al-Suhrawardy. Publisher: Carol Publishing Group.

"WHAT'S RIGHT WITH ISLAM: A NEW VISION FOR MUSLIMS AND THE WEST." by Feisal Abdul Rauf. Publisher: HarperCollins Publishers.

"ON BEING A MUSLIM" by Farid Esack. Publisher: Oneworld Publications, Oxford, U.K, 1999.

"ISLAM IN FOCUS" by Hammudah Abdalati. Publisher: American Trust Publications. Indianapolis.

"UNDERSTANDING ISLAM" by Yahiya Emerick. Publisher Alpha Books, Indianapolis.

"THE COMPLETE IDIOT'S GUIDE(R) TO UNDERSTANDING ISLAM" by Qasim Najar, Yahiya John Emerick. Publisher: Alpha Books,

"ISLAMIC SPIRITUALITY FOUNDATIONS" edited by Sayyed Hossein Nasr. Publisher: The Crossroad Publishing Company, New York.

"THE VISION OF ISLAM" by Sachiko Murata and William C. Chittick. Publisher: Paragon House, New York.

"THE GOSPEL OF ISLAM" by Duncan Greenlees. Publisher: Theosophical Publishing House, Post Box 270, Wheaton, Illinois 60187.

"ISLAMIC CONCEPT OF GOD" by Muhammad Zia Ullah. Publisher: KPI Limited, Boston.

"ISLAM: A CHALLENGE TO RELIGION" By G.A.Parwez. Publisher: Islamic Book Service, New Delhi. 11002, India.

HISTORY OF ISLAM

"Muhammad In Europe" by Minou Reeves. Publisher: New York University Press, New York.

"Muhammad: A Biography of the Prophet" by Karen Armstrong. Publisher: HarperCollins Publishers, New York.

"Muhammad Rasulallah" by Abul Hasan Ali Nadwi. Publisher: Academy of Islamic Rsearch and Publication. Targore Marg (Nadwa), P.O.Box 119, Lucknow, India.

"History of God" by Karen Armstrong. Publisher: Ballantine Books, New York.

"Image of the Prophet Muhammad in the west" by Jabal Muhammad Busben. Publisher: Islamic Foundation, Leicester, United Kingdom.

INDEX

Printed in the United States
35611LVS00002B/82

9 781413 465808